An analytic semigroup approach to convolution Volterra equations

An analytic semigroup approach to convolution Volterra equations

Proefschrift

ter verkrijging van de graad van doctor
aan de Technische Universiteit Delft,
op gezag van de Rector Magnificus prof.dr.ir. J.T. Fokkema,
voorzitter van het College voor Promoties,
in het openbaar te verdedigen op dinsdag 15 april 2003 om 13.30 uur

door

Kristina Willemina HOMAN

doctorandus in de wiskunde,
geboren te Groningen.

Dit proefschrift is goedgekeurd door de promotor:
Prof.dr. Ph.P.J.E. Clément

Samenstelling promotiecommissie:

Rector Magnificus,	voorzitter
Prof.dr. Ph.P.J.E. Clément,	Technische Universiteit Delft, promotor
Prof.dr. G. Da Prato,	Scuola Normale Superiore di Pisa, Italië
Dr. G.W. Desch,	Karl-Franzens-Universität Graz, Oostenrijk
Prof.dr.ir. A.W. Heemink,	Technische Universiteit Delft
Prof.dr. W.Th.F. den Hollander,	Technische Universiteit Eindhoven, Eurandom, Eindhoven
Prof.dr. S.-O. Londen,	Helsinki University of Technology, Finland
Dr. J.M.A.M. van Neerven,	Technische Universiteit Delft

Dr. G.W. Desch heeft in belangrijke mate aan de totstandkoming van het proefschrift bijgedragen.

Het onderzoek beschreven in dit proefschrift is mede gefinancierd door de Nederlandse Organisatie voor Wetenschappelijk Onderzoek (NWO) en de Stichting Wiskunde-onderzoek Nederland (SWON), projectnummer 613-06-558. Het Stieltjes Instituut heeft bijgedragen in de drukkosten van het proefschrift.

THOMAS STIELTJES INSTITUTE FOR MATHEMATICS

Published and distributed by: DUP Science

DUP Science is an imprint of Delft University Press

P.O. Box 98	Telephone:	+31 15 27 85 678
2600 MG Delft	Telefax:	+31 15 27 85 706
The Netherlands	E-mail:	Info@Library.TUDelft.NL

ISBN 90-407-2396-6

Keywords: Volterra equations, analytic semigroups, interpolation theory

Cover illustration: `http://www.terragalleria.com`. 'Ice and reflexions', McBride inlet, Glacier Bay National Park, Alaska. Copyright of Quang-Tuan Luong; reproduced with his permission. Design by Matthijs Klinkert.

Printed in The Netherlands

Opgedragen aan mijn ouders

Dag is nooit zo nat,
of zun schient aaltied wat.

Voorwoord

Aan het einde van mijn promotietijd bij de leerstoel Functionaalanalyse van de Technische Universiteit Delft wil ik bovenal mijn promotor Philippe Clément bedanken. Ik heb grote achting voor de inspirerende en motiverende manier waarop Philippe mij wiskunde heeft geleerd en ik beschouw het als een voorrecht om één van zijn promovendi te zijn.
Daarnaast ben ik veel dank verschuldigd aan Wolfgang Desch van de Karl-Franzens-Universität Graz. Zonder de prettige samenwerking met Wolfgang zou de inhoud van dit proefschrift niet tot stand zijn gekomen. Zijn inzichten zijn zeer leerzaam voor me geweest.
Ik wil mijn collega's Erik Koelink, Jan van Neerven, Ben de Pagter en Guido Sweers bedanken voor de plezierige en open omgeving die zij voor mij waren. Mijn kamergenoten Carolus Reinecke, Henrico Witvliet, Yvette van Norden en Anna Dall'Acqua dank ik voor de leuke werktijd samen. Mijn dank gaat ook uit naar Ingeborg Goddijn voor de vele keren lunch-zwemmen en naar Dori Steeneken voor de gezelligheid die ik altijd bij haar kon vinden.
Verder wil ik mijn vrienden, mijn familie en Ria en Wim bedanken voor de blijvende belangstelling voor mijn voortgang en welzijn. In het bijzonder noem ik mijn paranimfen Taco Homan en Sacha la Bastide-van Gemert.
Speciale dank gaat uit naar mijn ouders, die mij met hun begrip voor wiskundig onderzoek altijd door een moeilijke dag heen wisten te praten.
Tenslotte bedank ik Jorg, mijn gewezen voorman op de kapitalistische barricades, mijn wederhelft en mijn glimlach.

Krista Homan
's Gravenhage, februari 2003

Contents

Introduction

This thesis is concerned with Volterra integrodifferential equations of convolution type with completely monotonic kernels. The main objective is to provide an analytic semigroup setting for these equations, based on the complete monotonicity of the kernel. Bernstein's theorem allows for rewriting the Volterra equation into an abstract Cauchy problem in an appropriate infinite dimensional Hilbert space, in such a way that the operator governing this problem generates an analytic semigroup. Then the solution to the abstract Cauchy problem, together with interpolation methods, is used to obtain existence and regularity of solutions to the Volterra equation, as well as a representation formula.
The type of Volterra equations studied in this thesis is the semilinear X-valued Volterra integrodifferential equation

$$\begin{aligned} \frac{\mathrm{d}}{\mathrm{d}t}\int_{-\infty}^{t} a(t-s)u(s)\,\mathrm{d}s &= f(t,u(t)), \quad t>0, \\ u(t) &= u_0(t), \qquad\quad t\le 0, \end{aligned} \tag{1}$$

where X is a real separable Hilbert space, where the function $u_0 : (-\infty, 0] \to X$ is locally integrable, where the function $f : [0,\infty)\times X \to X$ is Hölder continuous in the first variable and globally Lipschitz continuous in the second, and where the kernel $a : (0,\infty) \to \mathbb{R}$ is completely monotonic, locally integrable, and singular at zero.
A motivation for studying this type of Volterra equations emanates from models involving memory, for example in the theory of viscoelasticity where the nonlinear stress-strain relations are of memory type, see [RHN87] and [Eng96]. As another example we consider heat flow in a material with memory as described in [Nun71] and [CDP97]. We also refer to [CN81], [Noh81], [LN84], [CP90], [Lun90], [Prü93], [GL95], and to [Jak01] and [Zac03].

Example. We consider heat flow in a bar of length l. At time $t \in \mathbb{R}$ and position $x \in [0,l]$ the temperature of the bar is denoted by $u(t,x)$, the density of internal energy by $e(t,x)$, the heat flux by $q(t,x)$, and the heat supply by $f(t,x)$. The functions e, q, and f are related to each other by the energy balance (the law of conservation of energy)

$$\tfrac{\partial}{\partial t}e(t,x) = -\tfrac{\partial}{\partial x}q(t,x) + f(t,x), \quad t \in \mathbb{R},\, x \in [0,l]. \tag{2}$$

Moreover, e and q are related to the temperature u and its gradient $\tfrac{\partial}{\partial x}u$ by the constitutive laws

$$e(t,x) = b_0 u(t,x) + \int_{-\infty}^{t} \beta(t-s)u(s,x)\,\mathrm{d}s, \quad t \in \mathbb{R},\, x \in [0,l], \tag{3}$$

$$q(t,x) = -c_0 \tfrac{\partial}{\partial x}u(t,x) + \int_{-\infty}^{t} \gamma(t-s)\tfrac{\partial}{\partial x}u(s,x)\,\mathrm{d}s, \quad t \in \mathbb{R},\, x \in [0,l], \tag{4}$$

where the constants $b_0 \geq 0$ and $c_0 \geq 0$ denote the heat capacity respectively the thermal conductivity of the bar, and where the functions β and γ denote the internal energy relaxation function respectively the heat flux relaxation function of the bar. By combining (2), (3), and (4) we obtain the linear heat equation with memory

$$\begin{aligned}&\frac{\partial}{\partial t}\left(b_0 u(t,x) + \int_{-\infty}^{t} \beta(t-s)u(s,x)\,\mathrm{d}s\right) \\ &\quad = c_0 \tfrac{\partial^2}{\partial x^2}u(t,x) - \textstyle\int_{-\infty}^{t} \gamma(t-s)\tfrac{\partial^2}{\partial x^2}u(s,x)\,\mathrm{d}s + f(t,x), \quad t \in \mathbb{R},\, x \in [0,l].\end{aligned} \tag{5}$$

The functions β and γ usually are of the form

$$\sum_{k=1}^{n} c_k \mathrm{e}^{-\alpha_k t}, \quad t > 0,$$

where $n \in \mathbb{N}$ and c_k, $\alpha_k > 0$ for $k = 1, \cdots, n$, or more generally, β and γ are completely monotonic functions. For the sake of simplicity we assume that γ is identically zero.

In order to characterize u completely we need initial and boundary value conditions. We choose Dirichlet boundary conditions

$$u(t,0) = u(t,l) = 0, \quad t \geq 0. \tag{6}$$

In addition, we assume that at every position $x \in [0,l]$ the history of the temperature $u(t,x)$ is known at every time $t \leq 0$, that is,

$$u(t,x) = u_0(t,x), \quad t \leq 0,\, x \in [0,l], \tag{7}$$

for some given function u_0.

We reformulate the linear heat equation (5) as a Volterra integrodifferential equation in a Hilbert space X by means of an unbounded linear operator $L : \mathrm{D}(L) \subseteq X \to X$. Let $X := L^2(0,l)$ and let L be defined by

$$\mathrm{D}(L) := W^{2,2}(0,l) \cap W_0^{1,2}(0,l),$$

$$(Lu)(x) := \tfrac{\mathrm{d}^2}{\mathrm{d}x^2} u(x), \quad u \in \mathrm{D}(L),\ x \in (0,l).$$

Now (5) together with (6) and (7) becomes

$$\begin{aligned} \frac{\mathrm{d}}{\mathrm{d}t}\left(b_0 u(t) + \int_{-\infty}^{t} \beta(t-s) u(s)\,\mathrm{d}s \right) &= c_0 L u(t) + f(t), \quad && t > 0, \\ u(t) &= u_0, && t \leq 0. \end{aligned} \tag{8}$$

Note that when β is identically zero, the Volterra integrodifferential equation (8) reduces to an ordinary differential equation in X. We are interested in the case $b_0 = 0$. In this case we can replace the unbounded linear operator L in (8) by its Yosida approximation, which is a bounded linear operator, so that problem (8) is of the form (1), and then pass to the limit. However, this is beyond the scope of the thesis. □

The semigroup approach based on the complete monotonicity of the kernel has been initiated in [DM88], has been used in the context of viscoelasticity in [DG89], and has been generalized to a larger class of kernels in [Sta94]. In contrast to other semigroup approaches to Volterra equations, as for example in [Mil74] and [GLS90], this one leads to an analytic semigroup. The price for embedding integral or delay equations in a semigroup setting often is some loss of regularity. However, this approach yields regularity results which are close to optimal.

We are able to treat existence and regularity of solutions to problem (1) within the setting of analytic semigroups and interpolation spaces, see [BL76], [Tri78], [Paz83], [CHA+87], [Ama95], [Lun95], and [MCSA01]. This sheds a new light on results obtained by more direct methods, for which we refer to [DPI85], [GLS90], [GV91], [Prü93], and [Baj01]. Moreover, it opens access to a powerful analytic machinery, in particular tools for a stochastic perturbation of problem (1) by white noise. Finally, this setting paves the way for a formulation of the Kolmogorov equations corresponding to a stochastic perturbation of (1). For the well-established theory of stochastic perturbation in combination with analytic semigroups and Kolmogorov equations we refer to [DPZ92]. Furthermore, we refer to [DPZ96], [CDPP97], [BvN00], [KZ00a], [KZ00b], [Bon01], [BT01], and [DPZ02].

For the moment we restrict ourselves to the scalar linear Volterra integrodifferential equation

$$\begin{aligned} \frac{\mathrm{d}}{\mathrm{d}t}\int_{-\infty}^{t} a(t-s)u(s)\,\mathrm{d}s = f(t), \quad t>0, \\ u(t) = u_0(t), \quad t \leq 0, \end{aligned} \tag{9}$$

where the functions $u_0 : (-\infty, 0] \to \mathbb{R}$ and $f : [0, \infty) \to \mathbb{R}$ are locally integrable, and where the kernel $a : (0, \infty) \to \mathbb{R}$ is completely monotonic, locally integrable, and singular at zero. In case u_0 is identically zero it is known that problem (9) admits a unique solution u, and that u has the representation

$$u(t) = \begin{cases} \int_0^t b(t-s)f(s)\,\mathrm{d}s, & t>0, \\ 0, & t \leq 0, \end{cases} \tag{10}$$

where $b : (0, \infty) \to \mathbb{R}$ is the resolvent of the first kind of the kernel a, that is, b has the same properties as a and satisfies

$$\int_0^t a(t-s)b(s)\,\mathrm{d}s = 1, \quad t>0,$$

see [GLS90]. As our model problem we choose the kernel a to be given by

$$a(t) := \frac{t^{-\alpha}}{\Gamma(1-\alpha)}, \quad t>0,$$

where $\alpha \in (0, 1)$. Then the Volterra equation (9) amounts to the fractional derivative of the function u of order α:

$$\begin{aligned} \mathrm{D}^{\alpha}\, u(t) &= f(t), \quad t>0, \\ u(t) &= 0, \qquad t \leq 0. \end{aligned} \tag{11}$$

Even more, the resolvent b of the first kind of a is given by

$$b(t) := \frac{t^{1-\alpha}}{\Gamma(\alpha)}, \quad t>0,$$

and the solution u given by (10) is the fractional integral of the function f of order α.

In this thesis we extend the result above to the case when u_0 is not identically zero, for which to our knowledge there is no general theory. We always assume that a belongs to the class $\mathcal{K}$ of completely monotonic kernels, defined by

$$\mathcal{K} := \big\{a : (0, \infty) \to \mathbb{R};\ a \text{ completely monotonic},\ a \in L^1(0, 1), a(0+) = +\infty\big\}.$$

With every kernel $a \in \mathcal{K}$ we associate a number $\alpha(a) \in [0,1]$ that plays the role of α in the case of our model problem (11).

Heuristically, the semigroup approach can be explained as follows. By Bernstein's theorem, see [Wid41], there exists a unique Borel measure ν on $[0,\infty)$ such that

$$a(t) = \int_{[0,\infty)} \mathrm{e}^{-\kappa t}\, \nu(\mathrm{d}\kappa), \quad t > 0. \tag{12}$$

The idea is to substitute this expression into (9). For this purpose we define the function $\psi : [0,\infty) \times [0,\infty) \to \mathbb{R}$ by

$$\psi(t,\kappa) := \int_{-\infty}^{t} \mathrm{e}^{-\kappa(t-s)} u(s)\,\mathrm{d}s, \quad t \geq 0,\ \kappa \geq 0,$$

where the function $u : \mathbb{R} \to \mathbb{R}$ is such that (9) holds. Note that ψ is the solution to the initial value problem

$$\begin{aligned} \tfrac{\partial}{\partial t}\psi(t,\kappa) &= u(t) - \kappa\psi(t,\kappa), \quad t > 0,\ \kappa \geq 0, \\ \psi(0,\kappa) &= \psi_0(\kappa), \qquad\qquad\qquad\quad \kappa \geq 0, \end{aligned} \tag{13}$$

where the initial value $\psi_0 : [0,\infty) \to \mathbb{R}$ is defined by

$$\psi_0(\kappa) := \int_0^{\infty} \mathrm{e}^{-\kappa s} u_0(-s)\,\mathrm{d}s, \quad \kappa \geq 0.$$

By substituting (12) into (9) we obtain

$$\begin{aligned} \frac{\mathrm{d}}{\mathrm{d}t}\int_{-\infty}^{t} a(t-s)u(s)\,\mathrm{d}s &= \frac{\mathrm{d}}{\mathrm{d}t}\int_{-\infty}^{t}\left(\int_{[0,\infty)} \mathrm{e}^{-\kappa(t-s)}\,\nu(\mathrm{d}\kappa)\right) u(s)\,\mathrm{d}s \\ &= \frac{\mathrm{d}}{\mathrm{d}t}\int_{[0,\infty)}\left(\int_{-\infty}^{t} \mathrm{e}^{-\kappa(t-s)} u(s)\,\mathrm{d}s\right)\nu(\mathrm{d}\kappa) = \frac{\mathrm{d}}{\mathrm{d}t}\int_{[0,\infty)} \psi(t,\kappa)\,\nu(\mathrm{d}\kappa) \\ &= \int_{[0,\infty)} \tfrac{\partial}{\partial t}\psi(t,\kappa)\,\nu(\mathrm{d}\kappa) = \int_{[0,\infty)} \big(u(t) - \kappa\psi(t,\kappa)\big)\,\nu(\mathrm{d}\kappa), \qquad t > 0. \end{aligned}$$

Therefore, ψ is subject to the constraint

$$\int_{[0,\infty)} \big(u(t) - \kappa\psi(t,\kappa)\big)\,\nu(\mathrm{d}\kappa) = f(t), \quad t > 0. \tag{14}$$

We remark that at this stage we have a differential equation without memory, namely (13). In order to solve the initial value problem (13) we use a semigroup approach to rewrite (13) together with the constraint (14) into an abstract differential equation in a suitable Hilbert space. We proceed in two steps.

The first step is to assume that f is identically zero. In this case it appears that we can rewrite the initial value problem (13) and the constraint (14) into the homogeneous abstract Cauchy problem

$$\begin{aligned} \tfrac{\mathrm{d}}{\mathrm{d}t}\varphi(t) &= A\varphi(t), \quad t > 0, \\ \varphi(0) &= \psi_0, \end{aligned} \tag{15}$$

where A is the infinitesimal generator of a quasi-contractive analytic semigroup $\{S(t)\}_{t\geq 0}$ on a Hilbert space H. In this way we obtain a solution φ to the homogeneous abstract Cauchy problem (15). It remains to find a way going from φ to a solution u to problem (9) with f identically zero. It turns out that this is possible by means of an unbounded linear functional J on H, so that u has the representation

$$u(t) = \begin{cases} J(\varphi(t)), & t > 0, \\ u_0(t), & t \leq 0. \end{cases}$$

We fix an arbitrary $\beta > 0$ and define the Hilbert space H of equivalence classes by

$$H := \Big\{\varphi : [0,\infty) \to \mathbb{R}; \varphi \text{ Borel measurable and} \\ \int_{[0,\infty)} |\varphi(\kappa)|^2(\kappa+\beta)\,\nu(\mathrm{d}\kappa) < \infty\Big\}.$$

We want A to act on ψ like

$$\big(A\psi(t,\cdot)\big)(\kappa) = u(t) - \kappa\psi(t,\kappa), \quad t > 0,\ \kappa \geq 0, \tag{16}$$

with the additional condition that

$$\int_{[0,\infty)} \big(u(t) - \kappa\psi(t,\kappa)\big)\,\nu(\mathrm{d}\kappa) = 0, \qquad t > 0.$$

Therefore we define the linear functional $J : \mathrm{D}(J) \subseteq H \to \mathbb{R}$ by

$$\mathrm{D}(J) := \big\{\varphi \in H; \text{ there exists (a unique) } u \in \mathbb{R} \text{ such that } \kappa \mapsto u - \kappa\varphi(\kappa) \in H\big\},$$

$$J(\varphi) := u, \quad \varphi \in \mathrm{D}(J).$$

Now we define A as follows:

$$\mathrm{D}(A) := \left\{\varphi \in \mathrm{D}(J); \int_{[0,\infty)} \big(J(\varphi) - \kappa\varphi(\kappa)\big)\,\nu(\mathrm{d}\kappa) = 0\right\},$$

$$(A\varphi)(\kappa) := J(\varphi) - \kappa\varphi(\kappa), \quad \varphi \in \mathrm{D}(A),\ \kappa \geq 0.$$

In the second step we continue with f not identically zero. Without loss of generality we assume that u_0 is identically zero. Let the linear functional $I : \mathrm{D}(J) \to \mathbb{R}$ be defined by

$$I(\varphi) := \int_{[0,\infty)} \big(J(\varphi) - \kappa\varphi(\kappa)\big)\, \nu(\mathrm{d}\kappa), \quad \varphi \in \mathrm{D}(J).$$

Since the constraint (14) suggests that $I\big(\psi(t,\cdot)\big) = f(t)$ for every $t > 0$, the mapping $\kappa \mapsto \psi(t,\kappa)$ does not belong to $\mathrm{D}(A)$ for every $t > 0$. In particular, (16) will not hold. Therefore the homogeneous abstract Cauchy problem (15) is of no use. The idea is to modify (15). For this purpose we think of a function $\pi \in \mathrm{D}(J)$ with $I(\pi) = 1$, that is, π has the property that for every $\varphi \in \mathrm{D}(J)$ the function $\varphi - I(\varphi)\pi$ belongs to $\mathrm{D}(A)$. It appears that the following choice is satisfactory:

$$\pi(\kappa) := \frac{1}{\beta\hat{a}(\beta)} \frac{1}{\kappa + \beta}, \quad \kappa \geq 0.$$

Furthermore, we define the linear operator $\tilde{A} : \mathrm{D}(J) \to H$ by

$$(\tilde{A}\varphi)(\kappa) := J(\varphi) - \kappa\varphi(\kappa), \quad \varphi \in \mathrm{D}(J),\ \kappa \geq 0.$$

It turns out that $\tilde{A}(\pi) = \beta\pi$. Now we observe that if $J\big(\psi(t,\cdot)\big) = u(t)$ for every $t > 0$, then we have

$$\begin{aligned} \tfrac{\mathrm{d}}{\mathrm{d}t}\psi(t,\cdot) &= \tilde{A}\psi(t,\cdot) = A\Big(\psi(t,\cdot) - I\big(\psi(t,\cdot)\big)\pi\Big) + I\big(\psi(t,\cdot)\big)\tilde{A}\pi \\ &= A\big(\psi(t,\cdot) - f(t)\pi\big) + \beta f(t)\pi, \quad t > 0. \end{aligned}$$

This suggests to consider the following inhomogeneous abstract Cauchy problem in an extrapolation space H_{-1} of H:

$$\begin{aligned} &\tfrac{\mathrm{d}}{\mathrm{d}t}\varphi(t) = A_{-1}\varphi(t) + f(t)(\beta I_H - A_{-1})\pi, \quad t > 0, \\ &\varphi(0) = 0. \end{aligned} \tag{17}$$

However, if φ is a solution to problem (17), then we cannot expect φ to belong to $\mathrm{D}(J)$, since $(\beta I_H - A_{-1})\pi$ belongs to H_{-1}. Therefore we would like π to belong to an interpolation space $(H, \mathrm{D}(A))_{\theta,2}$ and J to have a continuous extension to an interpolation space $(H, \mathrm{D}(A))_{\eta,2}$, for some $\theta, \eta \in (0,1)$ with $\eta \leq \theta$. Such η and θ exist indeed. In this way we obtain a solution φ to the inhomogeneous abstract Cauchy problem (17) with values in $(H, \mathrm{D}(A))_{\eta,2}$. Moreover, problem (9) with u_0 identically zero admits a solution u, and u has the representation

$$u(t) = \begin{cases} J(\varphi(t)), & t > 0, \\ 0, & t \leq 0. \end{cases}$$

One aspect of our semigroup approach is not immediately clear. A solution φ to the homogeneous abstract Cauchy problem (15) is an equivalence class in the Hilbert space, but we shall need a version of the equivalence class φ to show that the constructed function u is a solution to problem (9). This problem is considered in Chapter 2. The main result is that if the semigroup is analytic, then the solution φ to the homogeneous abstract Cauchy problem has an analytic version.

Chapter 3 is concerned with the scalar linear Volterra integrodifferential equation (9). This chapter is the core of the thesis, where the semigroup approach as explained above is done in detail. In Chapter 4 we study three different types of Volterra integrodifferential equations in a separable Hilbert space X. We use the X-valued equivalents of the Hilbert space H, the linear functional J, the linear operator A, and the analytic semigroup $\{S(t)\}_{t\geq 0}$. These are denoted by respectively $\mathcal{H}$, $\mathcal{J}$, $\mathcal{A}$, and $\{\mathcal{S}(t)\}_{t\geq 0}$. Section 4.1 treats the Hilbert-valued equivalent of the scalar linear Volterra integrodifferential equation (9), which is a straightforward generalization of the scalar case. We state the two main results in the Hilbert-valued case.

Hypothesis 4.1.3. The function $u_0 : (-\infty, 0] \to X$ is strongly Borel measurable and has the following properties:

(i) There exist $M_1 > 0$ and $\omega > 0$ such that $\|u_0(t)\| \leq M_1 e^{\omega t}$ for every $t \leq 0$;

(ii) There exist $M_2 > 0$ and $\delta > 0$ such that $\|u_0(0) - u_0(t)\| \leq M_2 |t|$ for every $t \in [-\delta, 0]$.

Theorem 4.1.20. Let u_0 satisfy Hypothesis 4.1.3. If a belongs to $\mathcal{K}$ with $\alpha(a) \in (0,1)$, then the X-valued equivalent of problem (9) with f identically zero admits a unique solution u. Moreover, u is real analytic in $(0,\infty)$, belongs to $C^{0,\zeta}([0,T];X)$ for all $T > 0$ and $\zeta \in [0, \alpha(a))$, and has the representation

$$u(t) = \begin{cases} \mathcal{J}\big(\mathcal{S}(t)\psi_0\big), & t > 0, \\ u_0(t), & t \leq 0. \end{cases}$$

Theorem 4.1.26. Let $f : [0,\infty) \to X$ belong to $L^p(0,T;X)$ for some $p \in [1,\infty)$ and every $T > 0$. If a belongs to $\mathcal{K}$ with $\alpha(a) \in (0,1)$, then the X-valued equivalent of problem (9) with u_0 identically zero admits a unique solution u, and u has the representation

$$u(t) = \begin{cases} \int_0^t J\big((\beta I_H - A)S(t-s)\pi\big) f(s)\,\mathrm{d}s, & t > 0, \\ 0, & t \leq 0. \end{cases}$$

Moreover, for all $T > 0$ and $\zeta \in (0, \alpha(a))$ the following holds:

(i) If $p = 1$, then u belongs to $L^q(0,T;X)$ for every $q \in \big[1, \frac{1}{1-\zeta}\big)$;

(ii) If $p \in \big(1, \frac{1}{\zeta}\big)$, then u belongs to $L^q(0,T;X)$ where $q := \frac{p}{1-\zeta p}$;

(iii) If $p = \frac{1}{\zeta}$, then u belongs to $L^q(0,T;X)$ for every $q \in [1,\infty)$;

(iv) If $p \in \big(\frac{1}{\zeta}, \infty\big)$, then u belongs to $C_0^{0,\zeta-\frac{1}{p}}([0,T];X)$.

In particular, if f belongs to $C_0^{0,\gamma}([0,T];X)$ for some $\gamma \in (0,1)$ and every $T > 0$, then u belongs to

$$C^{[\gamma+\zeta],\gamma+\zeta-[\gamma+\zeta]}([0,T];X)$$

for all $T > 0$ and $\zeta \in (0, \alpha(a))$ such that $\zeta \neq 1-\gamma$.

Note that Theorem 4.1.26 is an extension of the Hardy-Littlewood inequalities concerning fractional integrals, see [HL28] and [GV91].
As a consequence of our semigroup approach we obtain a representation of the resolvent of the first kind of the kernel $a \in \mathcal{K}$.

Theorem 3.7.7. If a belongs to $\mathcal{K}$ with $\alpha(a) \in (0,1)$, then the resolvent b of the first kind of a has the representation

$$b(t) = J\big((\beta I_H - A)S(t)\pi\big), \quad t > 0.$$

If, in addition, $\alpha(a) > \frac{1}{2}$, then b belongs to $L^2(0,T)$ for every $T > 0$.

In Section 4.2 we discuss the semilinear Hilbert-valued Volterra integrodifferential equation (1). The X-valued equivalent of the abstract Cauchy problem (17) is the semilinear initial value problem

$$\begin{aligned} &\tfrac{\mathrm{d}}{\mathrm{d}t}\varphi(t) = \mathcal{A}_{-1}\varphi(t) + (\beta I_H - A_{-1})\pi \otimes f\big(t, \mathcal{J}(\varphi(t))\big), \quad t > 0, \\ &\varphi(0) = \psi_0, \end{aligned} \tag{18}$$

where a function $\varphi \otimes x$ in $\mathcal{H}$ for all $\varphi \in H$ and $x \in X$ is defined by

$$(\varphi \otimes x)(\kappa) := \varphi(\kappa)x, \quad \kappa \geq 0.$$

The main result of this section is the following:

Hypothesis 4.2.2. The function $f : [0,\infty) \times X \to X$ is such that

(i) For every $T > 0$ there exists $L > 0$ such that

$$\big\|f(t,x) - f(t,y)\big\| \leq L\|x - y\|, \quad t \in [0,T],\ x, y \in X;$$

(ii) There exists $\vartheta \in (0,1)$ such that for all $T > 0$ and $x_0 \in X$ there exist $r > 0$ and $K > 0$ such that

$$\big\|f(t,x) - f(s,x)\big\| \le K|t-s|^{\vartheta}, \quad s,t \in [0,T],\ x \in B_X(x_0;r).$$

Theorem 4.2.9. Let u_0 and f satisfy Hypothesis 4.1.3 respectively Hypothesis 4.2.2. If a belongs to $\mathcal{K}$ with $\alpha(a) \in (0,1)$, then problem (1) admits a unique solution u, and u has the representation

$$u(t) = \begin{cases} \mathcal{J}(\varphi(t)), & t > 0, \\ u_0(t), & t \le 0, \end{cases}$$

where φ is the strict solution to problem (18) according to [Lun95]. Moreover, u belongs to $C^{0,\zeta}([0,T];\,X)$ for all $T > 0$ and $\zeta \in (0,\alpha(a))$, and u satisfies

$$u(t) = \mathcal{J}\big(\mathcal{S}(t)\psi_0\big) + \int_0^t J\big((\beta I_H - A)S(t-s)\pi\big) f(s,u(s))\,\mathrm{d}s, \quad t \ge 0.$$

Motivated by the application of a stochastic perturbation of problem (9), Section 4.3 is concerned with the first kind linear Hilbert-valued Volterra equation

$$\int_0^t a(t-s)u(s)\,\mathrm{d}s = h(t), \quad t \ge 0, \tag{19}$$

where the function $h : [0,\infty) \to X$ belongs to $C_0([0,T];X)$ for every $T > 0$, see also [Gri80], [Gri85], [GLS90], and [GV91]. We obtain the next theorem:

Theorem 4.3.3. Let h belong to $C_0^{0,\gamma}([0,T];X)$ for some $\gamma \in (1-\alpha(a),1)$ and every $T > 0$. If a belongs to $\mathcal{K}$ with $\alpha(a) \in (0,1)$, then problem (19) admits a unique solution u. Moreover, u belongs to $C^{0,\gamma+\zeta-1}([0,T];X)$ for all $T > 0$ and $\zeta \in (1-\gamma,\alpha(a))$, and u has the representation

$$u(t) = \frac{\mathrm{d}}{\mathrm{d}t}\int_0^t J\big((\beta I_H - A)S(t-s)\pi\big)h(s)\,\mathrm{d}s, \quad t \ge 0.$$

Finally, Chapter 5 deals with the stochastic linear Volterra integrodifferential equation

$$\begin{aligned} \frac{\mathrm{d}}{\mathrm{d}t}\int_{-\infty}^t a(t-s)U(s)\,\mathrm{d}s &= f(t) + \sigma\dot{B}(t), \quad t > 0, \\ U(t) &= u_0(t), \qquad\qquad t \le 0, \end{aligned} \tag{20}$$

where $\sigma \in \mathbb{R}$, and where $\{B(t)\}_{t\ge 0}$ is standard Brownian motion. To give (20) a meaning we interpret it as the stochastic first kind linear $L^2(\Omega)$-valued Volterra equation

$$\int_0^t a(t-s)U(s)\,\mathrm{d}s = h(t)\mathbf{1}_\Omega + \sigma B(t), \quad t \ge 0, \tag{21}$$

where $h : [0, \infty) \to \mathbb{R}$ is given by

$$h(t) := \int_{-\infty}^{0} a(-s)u_0(s)\,\mathrm{d}s - \int_{-\infty}^{0} a(t-s)u_0(s)\,\mathrm{d}s + \int_0^t f(s)\,\mathrm{d}s, \quad t \geq 0.$$

In Section 5.2 we apply the results of Section 4.3 to problem (21). This leads to the following theorem:

Theorem 5.2.4. If a belongs to $\mathcal{K}$ with $\alpha(a) \in (\frac{1}{2}, 1)$, then problem (20), with f and u_0 identically zero and $\sigma = 1$, admits a unique solution U. Moreover, U belongs to $C^{0,\zeta}([0,T]; L^2(\Omega))$ for all $T > 0$ and $\zeta \in (0, \alpha(a) - \frac{1}{2})$, and U has the representation

$$U(t) = \begin{cases} \frac{\mathrm{d}}{\mathrm{d}t} \int_0^t J\big((\beta I_H - A)S(t-s)\pi\big)B(s)\,\mathrm{d}s, & t \geq 0, \\ 0, & t < 0. \end{cases}$$

In Section 5.3 we obtain that the solution U in Theorem 5.2.4 is a centered Gaussian system with covariance

$$\mathbb{E}\big(U(t)U(s)\big) = \int_0^{\min\{s,t\}} \big(J((\beta I_H - A)S(\tau)\pi)\big)^2\,\mathrm{d}\tau, \quad t, s \geq 0.$$

In particular, we prove that U is represented by the stochastic convolution

$$U(t) := \begin{cases} \int_0^t J\big((\beta I_H - A)S(t-s)\pi\big)\,\mathrm{d}B(s), & t \geq 0, \\ 0, & t < 0. \end{cases}$$

We conclude the introduction with the remark that the contents of Chapters 2 and 3, with the exception of Theorem 3.7.7, have appeared in [DH02] respectively [CDH02], and will appear in [DH] respectively [CDH].

Chapter 1

Preliminaries

1.1 Completely monotonic functions, Bernstein's theorem, and the Laplace transform

Definition 1.1.1 A function $f : (0, \infty) \to \mathbb{R}$ is called *completely monotonic* if f belongs to $C^\infty(0, \infty)$ and

$$(-1)^n \frac{\mathrm{d}^n}{\mathrm{d}x^n} f(x) \geq 0, \quad x > 0,\, n = 1, 2, \cdots .$$

Lemma 1.1.2 *A completely monotonic function $f : (0, \infty) \to \mathbb{R}$ has the following properties:*

(i) *If $f(x_0) = 0$ for some $x_0 > 0$, then f is identically zero;*

(ii) *The function f has an analytic extension to $\{z \in \mathbb{C};\ \mathrm{Re} z > 0\}$;*

(iii) *If $f(0+) = +\infty$, then $(-1)^n \frac{\mathrm{d}^n}{\mathrm{d}x^n} f(0+) = +\infty$ for $n = 1, 2, \cdots$;*

(iv) $(-1)^n \frac{\mathrm{d}^n}{\mathrm{d}x^n} f(+\infty) = 0$ *for $n = 1, 2, \cdots$.*

PROOF: For (i) and (ii) we refer to [GLS90, page 147, Theorem 2.8] respectively [Wid41, page 146, Theorem 3a]. For (iii) and (iv) we use that for all $0 < x < T$,

$$f(x) = f(T) + \int_x^T -\tfrac{\mathrm{d}}{\mathrm{d}y} f(y)\, \mathrm{d}y \leq f(T) - (T - x) \tfrac{\mathrm{d}}{\mathrm{d}x} f(x),$$

$$f(T) = f(x) + \int_x^T \tfrac{\mathrm{d}}{\mathrm{d}y} f(y)\, \mathrm{d}y \leq f(x) + (T - x) \tfrac{\mathrm{d}}{\mathrm{d}x} f(T).$$

□

For the following theorem we refer to [Wid41, pages 161, 12, Theorems 12b, 7a], [Tay65, page 219, Theorem 4-10I], and [WZ77, page 201, Theorem 11.11].

Theorem 1.1.3 (S. Bernstein) *A function $f : (0,\infty) \to \mathbb{R}$ is completely monotonic if and only if there exists a Borel measure ν on $[0,\infty)$ such that*

$$f(x) = \int_{[0,\infty)} \mathrm{e}^{-\kappa x}\, \nu(\mathrm{d}\kappa), \quad x > 0.$$

In this situation, the measure ν is uniquely determined by f.

Lemma 1.1.4 *Let $f : (0,\infty) \to \mathbb{R}$ be completely monotonic. Let ν be the unique Borel measure on $[0,\infty)$ such that $f(x) = \int_{[0,\infty)} \mathrm{e}^{-\kappa x}\, \nu(\mathrm{d}\kappa)$ for every $x > 0$. Then the following holds:*

(i) $\nu([0,T]) < \infty$ *for every* $T > 0$;

(ii) *It holds that*

$$(-1)^n \tfrac{\mathrm{d}^n}{\mathrm{d}x^n} f(x) = \int_{[0,\infty)} \mathrm{e}^{-\kappa x} \kappa^n\, \nu(\mathrm{d}\kappa), \quad x > 0,\ n = 0, 1, \cdots;$$

(iii) $f(0+) < \infty$ *if and only if* $\nu([0,\infty)) < \infty$, *and then* $f(0+) = \nu([0,\infty))$;

(iv) *If* $f(0+) = +\infty$, *then* $\int_{[0,\infty)} \kappa^n\, \nu(\mathrm{d}\kappa) = +\infty$ *for* $n = 0, 1, \cdots$;

(v) $f(+\infty) = \nu(\{0\})$.

PROOF: For (i) we refer to [Tay65, page 219, Theorem 4-10I]. □

Lemma 1.1.5 *Let $f : (0,\infty) \to \mathbb{R}$ be completely monotonic and belonging to $L^1(0,1)$. Then the Laplace transform $\hat{f} : (0,\infty) \to \mathbb{R}$ is completely monotonic. Moreover, if ν is the unique Borel measure on $[0,\infty)$ such that $f(x) = \int_{[0,\infty)} \mathrm{e}^{-\kappa x}\, \nu(\mathrm{d}\kappa)$ for every $x > 0$, then $\hat{f}$ has the following properties:*

(i) $\hat{f}(\lambda) = \int_{[0,\infty)} \frac{1}{\kappa+\lambda}\, \nu(\mathrm{d}\kappa)$ *for every* $\lambda > 0$;

(ii) $-\frac{\mathrm{d}}{\mathrm{d}\lambda}\hat{f}(\lambda) = \int_{[0,\infty)} \frac{1}{(\kappa+\lambda)^2}\, \nu(\mathrm{d}\kappa)$ *for every* $\lambda > 0$;

(iii) $\hat{f}(\lambda) + \lambda \frac{\mathrm{d}}{\mathrm{d}\lambda}\hat{f}(\lambda) = \int_{[0,\infty)} \frac{\kappa}{(\kappa+\lambda)^2}\, \nu(\mathrm{d}\kappa)$ *for every* $\lambda > 0$;

(iv) *The mapping* $\lambda \mapsto \lambda \hat{f}(\lambda)$ *is nondecreasing on* $(0,\infty)$;

(v) $\hat{f}(\lambda_0) - \hat{f}(\lambda) \le \frac{\lambda-\lambda_0}{\lambda_0} \hat{f}(\lambda)$ *for all* $0 < \lambda_0 < \lambda$;

(vi) $\left(1 - \frac{\lambda_0^2}{(\lambda-\lambda_0)^2}\right) \hat{f}(\lambda) + \frac{\lambda^2}{\lambda-\lambda_0} \frac{\mathrm{d}}{\mathrm{d}\lambda} \hat{f}(\lambda) + \frac{\lambda_0^2}{(\lambda-\lambda_0)^2} \hat{f}(\lambda_0) = \int_{[0,\infty)} \frac{\kappa^2}{(\kappa+\lambda)^2} \frac{1}{\kappa+\lambda_0}\, \nu(\mathrm{d}\kappa)$ *for all* $0 < \lambda_0 < \lambda$.

PROOF: The Laplace transform of f is well-defined since f is locally integrable and nonincreasing. $\square$

Lemma 1.1.6 *Let $f:(0,\infty)\to\mathbb{R}$ be completely monotonic and belonging to $L^1(0,1)$. Let ν be the unique Borel measure on $[0,\infty)$ such that $f(x)=\int_{[0,\infty)} \mathrm{e}^{-\kappa x}\,\nu(\mathrm{d}\kappa)$ for every $x>0$. Then the following holds:*

(i) $\int_{[T,\infty)} \frac{1}{\kappa}\,\nu(\mathrm{d}\kappa)<\infty$ *for every* $T>0$;

(ii) $\lim_{x\downarrow 0}(-1)^n x^{n+1}\frac{\mathrm{d}^n}{\mathrm{d}x^n} f(x)=0$ *for* $n=0,1,\cdots$.

PROOF: Property (i) follows from Lemma 1.1.5(i). Property (ii) is a result of Lemma 1.1.4(ii) and Lebesgue's dominated convergence theorem, since we have for $n=0,1,\cdots$ and every $x\in(0,1]$, using Lemmas 1.1.4(i) and 1.1.6(i),

$$(-1)^n x^{n+1}\tfrac{\mathrm{d}^n}{\mathrm{d}x^n} f(x)=\int_{[0,\infty)} \mathrm{e}^{-\kappa x}\kappa^n x^{n+1}\,\nu(\mathrm{d}\kappa)$$
$$\leq \int_{[0,1]} \nu(\mathrm{d}\kappa)+\max\{\mathrm{e}^{-t}t^{n+1};\, t\geq 0\}\int_{[1,\infty)} \frac{1}{\kappa}\,\nu(\mathrm{d}\kappa)<\infty.$$

$\square$

Lemma 1.1.7 *Let $f:(0,\infty)\to\mathbb{R}$ be completely monotonic and belonging to $L^1(0,1)$. Let ν be the unique Borel measure on $[0,\infty)$ such that $f(x)=\int_{[0,\infty)} \mathrm{e}^{-\kappa x}\,\nu(\mathrm{d}\kappa)$ for every $x>0$. Then the Laplace transform $\hat{f}:(0,\infty)\to\mathbb{R}$ has a continuous extension to $\{z\in\mathbb{C}\backslash\{0\};\ \mathrm{Re}z\geq 0\}$ and for all $\lambda\geq 0$ and $\omega\in\mathbb{R}$ such that $\lambda+\mathrm{i}\omega\neq 0$,*

$$\hat{f}(\lambda+\mathrm{i}\omega)=\int_{[0,\infty)} \frac{1}{\kappa+\lambda+\mathrm{i}\omega}\,\nu(\mathrm{d}\kappa)=\int_{[0,\infty)} \frac{\kappa+\lambda-\mathrm{i}\omega}{(\kappa+\lambda)^2+\omega^2}\,\nu(\mathrm{d}\kappa).$$

PROOF: By Lemma 1.1.2(ii) it is sufficient to prove that $\lim_{\lambda\downarrow 0}\hat{f}(\lambda+\mathrm{i}\omega)=\int_{[0,\infty)} \frac{1}{\kappa+\mathrm{i}\omega}\,\nu(\mathrm{d}\kappa)$ for every $\omega\neq 0$. Now this follows from Lebesgue's dominated convergence theorem, since we have using Lemmas 1.1.4(i) and 1.1.6(i),

$$\int_{[0,\infty)} \frac{1}{|\kappa+\mathrm{i}\omega|}\,\nu(\mathrm{d}\kappa)\leq \frac{1}{|\omega|}\int_{[0,1]} \nu(\mathrm{d}\kappa)+\int_{[1,\infty)} \frac{1}{\kappa}\,\nu(\mathrm{d}\kappa)<\infty,\quad \omega\neq 0.$$

$\square$

1.2 Titchmarsh's theorem

Theorem 1.2.1 (E.C. Titchmarsh) *If $f, g \in C[0,\infty)$ are such that*

$$\int_0^t f(t-s)g(s)\,\mathrm{d}s = 0, \quad t > 0, \tag{1.1}$$

then either f or g is identically zero.

For a proof of Titchmarsh's theorem we refer to [Yos95, page 166, Theorem].

Corollary 1.2.2 *Let $f, g : (0,\infty) \to \mathbb{R}$ be such that f and g belong to $L^1(0,T)$ for every $T > 0$. If (1.1) holds, then either $f(t) = 0$ for almost every $t > 0$ or $g(t) = 0$ for almost every $t > 0$.*

PROOF: Using the commutative and associative laws for convolutions it follows from (1.1) that the convolution of the mappings $t \mapsto \int_0^t f(s)\,\mathrm{d}s$ and $t \mapsto \int_0^t g(s)\,\mathrm{d}s$ is identically zero. Now Theorem 1.2.1 implies that either $\int_0^t f(s)\,\mathrm{d}s = 0$ for every $t \geq 0$ or $\int_0^t g(s)\,\mathrm{d}s = 0$ for every $t \geq 0$. Hence, either $f(t) = 0$ for almost every $t > 0$ or $g(t) = 0$ for almost every $t > 0$, see [Tay65, page 415, Lemma 9-8V]. □

1.3 Resolvents of the first kind

Definition 1.3.1 Let $a : (0,\infty) \to \mathbb{R}$ be such that a belongs to $L^1(0,T)$ for every $T > 0$. A *resolvent of the first kind of* a is a function $b : (0,\infty) \to \mathbb{R}$ such that b belongs to $L^1(0,T)$ for every $T > 0$ and

$$\int_0^t a(t-s)b(s)\,\mathrm{d}s = 1, \quad t > 0.$$

Theorem 1.3.2 *Let $a : (0,\infty) \to \mathbb{R}$ be such that a belongs to $L^1(0,T)$ for every $T > 0$. Then a has at most one resolvent of the first kind. If, in addition, a is completely monotonic and $a(0+) = +\infty$, then there exists a resolvent b of the first kind of a, and b is completely monotonic and $b(0+) = +\infty$.*

For the theorem above we refer to [GLS90, pages 158-159, Theorems 5.2, 5.4].

Example 1.3.3 The function $g_\alpha : (0,\infty) \to \mathbb{R}$ with $\alpha > 0$ is defined by

$$g_\alpha(t) := \frac{t^{\alpha-1}}{\Gamma(\alpha)}, \quad t > 0, \tag{1.2}$$

where we recall that the Gamma function $\Gamma : \{z \in \mathbb{C};\ \mathrm{Re} z > 0\} \to \mathbb{R}$ is the analytic function defined by

$$\Gamma(z) = \int_0^\infty e^{-t} t^{z-1}\,\mathrm{d}t, \quad \mathrm{Re} z > 0,$$

with the property that $\Gamma(z+1) = z\Gamma(z)$ whenever $\mathrm{Re} z > 0$, see [WW52, pages 235-241]. The function g_α has the following properties:

(i) $\hat{g}_\alpha(\lambda) = \lambda^{-\alpha}$ for every $\lambda > 0$;

(ii) For every $\gamma > 0$ it holds that

$$\int_0^t g_\alpha(t-s) g_\gamma(s)\,\mathrm{d}s = g_{\alpha+\gamma}(t), \quad t > 0;$$

(iii) If $\alpha < 1$, then g_α satisfies Bernstein's theorem with $\nu(\mathrm{d}\kappa) = g_{1-\alpha}(\kappa)\,\mathrm{d}\kappa$, that is,

$$g_\alpha(t) = \frac{1}{\Gamma(\alpha)\Gamma(1-\alpha)} \int_0^\infty e^{-\kappa t} \kappa^{-\alpha}\,\mathrm{d}\kappa, \quad t > 0.$$

Note that property (ii) shows that the resolvent of the first kind of g_α is $g_{1-\alpha}$.

For functions $f, g : (0,\infty) \to \mathbb{R}$ such that f and g belong to $L^1(0,T)$ for every $T > 0$ we denote the function $f * g : (0,\infty) \to \mathbb{R}$ by

$$(f * g)(t) := \int_0^t f(t-s) g(s)\,\mathrm{d}s, \quad t > 0.$$

Proposition 1.3.4 *Let $\alpha \in (0,1)$ and let f belong to $L^p(0,T)$ for some $p \in [1,\infty)$ and some $T > 0$. Then the following holds:*

(i) *If $p = 1$, then $g_\alpha * f$ belongs to $L^q(0,T)$ for every $q \in \left[1, \frac{1}{1-\alpha}\right)$ and there exists $c > 0$ depending on α and q such that*

$$\|g_\alpha * f\|_{L^q(0,T)} \le c\|f\|_{L^p(0,T)}; \tag{1.3}$$

(ii) *If $p \in \left(1, \frac{1}{\alpha}\right)$, then $g_\alpha * f$ belongs to $L^q(0,T)$ where $q := \frac{p}{1-\alpha p}$, and there exists $c > 0$ depending on α and p such that (1.3) holds;*

(iii) *If $p = \frac{1}{\alpha}$, then $g_\alpha * f$ belongs to $L^q(0,T)$ for every $q \in [1,\infty)$ and there exists $c > 0$ depending on α and q such that (1.3) holds;*

(iv) *If* $p \in \left(\frac{1}{\alpha}, \infty\right)$, *then* $g_\alpha * f$ *belongs to* $C_0^{0,\alpha-\frac{1}{p}}[0,T]$ *and there exists* $c > 0$ *depending on* α *and* p *such that*

$$\|g_\alpha * f\|_{L^\infty(0,T)} + T^{\alpha-\frac{1}{p}}\left[g_\alpha * f\right]_{\alpha-\frac{1}{p}} \leq cT^{\alpha-\frac{1}{p}}\|f\|_{L^p(0,T)},$$

where

$$\left[g_\alpha * f\right]_{\alpha-\frac{1}{p}} := \sup\left\{\frac{|(g_\alpha * f)(t) - (g_\alpha * f)(s)|}{(t-s)^{\alpha-\frac{1}{p}}};\ 0 \leq s < t \leq T\right\}.$$

PROOF: For properties (i), (ii), and (iii) we refer to [GV91, page 65, Theorems 4.1.1, 4.1.3] and [HL28]. For property (iv) let $p' \in (1,\infty)$ be such that $\frac{1}{p} + \frac{1}{p'} = 1$. Since $(\alpha - 1)p' > -1$ it follows that g_α belongs to $L^{p'}(0,T)$ and hence, $g_\alpha * f$ belongs to $L^1(0,T)$, is continuous on $[0,T]$, and $(g_\alpha * f)(0) = 0$, see [GLS90, page 40, Corollary 2.3(iii)]. For the rest of the result we refer to [GV91, page 67, Theorem 4.1.4]. □

1.4 Analytic semigroups and interpolation spaces

Let $(X, \|\cdot\|_X)$ be a Banach space.

Definition 1.4.1 A *strongly continuous semigroup of bounded linear operators on* X is a family $\{S(t)\}_{t\geq 0}$ in $\mathcal{L}(X)$ with the following properties:

(i) $S(0) = I_X$;

(ii) $S(t+s) = S(t)S(s)$ for all $t, s \geq 0$;

(iii) $\lim_{t\downarrow 0} S(t)x = x$ for every $x \in X$.

A strongly continuous semigroup $\{S(t)\}_{t\geq 0}$ is called *uniformly bounded* if there exists $M \geq 1$ such that $\|S(t)\|_{\mathcal{L}(X)} \leq M$ for every $t \geq 0$. In particular, if $M = 1$, then $\{S(t)\}_{t\geq 0}$ is called a *strongly continuous semigroup of contractions on* X.

Definition 1.4.2 The *infinitesimal generator* of a strongly continuous semigroup $\{S(t)\}_{t\geq 0}$ on X is the linear operator $A : \mathrm{D}(A) \subseteq X \to X$ defined by

$$\mathrm{D}(A) := \left\{x \in X;\ \lim_{t\downarrow 0}\frac{S(t)x - x}{t} \text{ exists}\right\},$$

$$Ax := \lim_{t\downarrow 0}\frac{S(t)x - x}{t}, \quad x \in \mathrm{D}(A).$$

Definition 1.4.3 Let $A : \mathrm{D}(A) \subseteq X \to X$ be a linear operator.

(i) A is called *dissipative in* $(X, \|\cdot\|_X)$ if $\|(\lambda I_X - A)x\|_X \geq \lambda \|x\|_X$ for all $x \in \mathrm{D}(A)$ and $\lambda > 0$;

(ii) A is called *m-dissipative in* $(X, \|\cdot\|_X)$ if A is dissipative in $(X, \|\cdot\|_X)$ and $\mathrm{Ran}(I_X - A) = X$.

A characterization of dissipative operators in Hilbert spaces is given in the next lemma. The proof is elementary, compare [Paz83, page 14, Theorem 4.2].

Lemma 1.4.4 *Let $(H, \langle\cdot,\cdot\rangle_H)$ be a real Hilbert space and $A : \mathrm{D}(A) \subseteq H \to H$ a linear operator. Then A is dissipative in $(H, \langle\cdot,\cdot\rangle_H)$ if and only if $\langle Ax, x\rangle_H \leq 0$ for every $x \in \mathrm{D}(A)$.*

Theorem 1.4.5 (Hille-Yosida) *A densely defined linear operator $A : \mathrm{D}(A) \subseteq X \to X$ is the infinitesimal generator of a strongly continuous semigroup of contractions on X if and only if A is m-dissipative in $(X, \|\cdot\|_X)$.*

For a proof of the Hille-Yosida theorem we refer to [Paz83, page 8, Theorem 3.1]. The following proposition gives an identification of the infinitesimal generator of a uniformly bounded semigroup. A proof can be found in [CHA$^+$87, page 47, Theorem 3.1(ii)].

Proposition 1.4.6 *Let $\{S(t)\}_{t\geq 0}$ be a uniformly bounded semigroup on X with infinitesimal generator A. Then*

$$(\lambda I_X - A)^{-1}x = \int_0^\infty \mathrm{e}^{-\lambda t} S(t)x \,\mathrm{d}s, \quad \lambda > 0,\, x \in X,$$

where the integral is Bochner in X.

Proposition 1.4.7 *Let $A : \mathrm{D}(A) \subseteq X \to X$ be an m-dissipative operator in $(X, \|\cdot\|_X)$. If X is reflexive, then A is densely defined in X.*

The proposition above is proved in [Paz83, page 16, Theorem 4.6]. The next result is proved in [Paz83, page 9, Lemma 3.2].

Lemma 1.4.8 *Let $A : \mathrm{D}(A) \subseteq X \to X$ be the infinitesimal generator of a strongly continuous semigroup on X. Then*

$$\lim_{\lambda\to\infty} \lambda(\lambda I_X - A)^{-1}x = x, \quad x \in X.$$

The following perturbation result is proved in [Paz83, page 76, Theorem 1.1].

Proposition 1.4.9 *Let $\{S(t)\}_{t\geq 0}$ be a strongly continuous semigroup on X with infinitesimal generator A and, for some $M \geq 1$ and $\omega \geq 0$, $\|S(t)\|_{\mathcal{L}(X)} \leq M\mathrm{e}^{\omega t}$ for every $t \geq 0$. If B belongs to $\mathcal{L}(X)$, then $A+B$ is the infinitesimal generator of a strongly continuous semigroup $\{T(t)\}_{t\geq 0}$ on X, satisfying*

$$\|T(t)\|_{\mathcal{L}(X)} \leq M\mathrm{e}^{(\omega+M\|B\|_{\mathcal{L}(X)})t}, \quad t \geq 0.$$

In particular, if $B = \lambda I_X$ for some $\lambda \in \mathbb{R}$, then $T(t) = \mathrm{e}^{\lambda t}S(t)$ for every $t \geq 0$.

Let $\Sigma_\delta \subseteq \mathbb{C}$ with $\delta \in (0,\pi]$ denote the sector $\Sigma_\delta := \{z \in \mathbb{C}\backslash\{0\}; |\operatorname{Arg}(z)| < \delta\}$, where $\operatorname{Arg}(z)$ is taken in $(\pi,\pi]$.

Definition 1.4.10 An *analytic semigroup in* Σ_δ is a family $\{S(z)\}_{z\in\Sigma_\delta\cup\{0\}}$ in $\mathcal{L}(X)$ with the following properties:

(i) $S(0) = I_X$;

(ii) $S(z_1+z_2) = S(z_1)S(z_2)$ for all $z_1, z_2 \in \Sigma_\delta$;

(iii) $\lim_{\substack{z\to 0\\ z\in\Sigma_\delta}} S(z)x = x$ for every $x \in X$;

(iv) The mapping $z \mapsto S(z)x$ is analytic in Σ_δ for every $x \in X$.

A strongly continuous semigroup $\{S(t)\}_{t\geq 0}$ on X is called *analytic* if it can be extended to an analytic semigroup $\{S(z)\}_{z\in\Sigma_\delta\cup\{0\}}$ in some sector Σ_δ with $\delta \in (0,\pi]$.

Theorem 1.4.11 *Let $\{S(t)\}_{t\geq 0}$ be a uniformly bounded, strongly continuous semigroup on X with infinitesimal generator A. If $0 \in \rho(A)$, then the following statements are equivalent:*

(i) *The semigroup can be extended to an analytic semigroup $\{S(z)\}_{z\in\Sigma_\delta\cup\{0\}}$ in a sector Σ_δ for some $\delta \in (0,\pi]$, and $\sup\{\|S(z)\|_{\mathcal{L}(X)};\, z \in \overline{\Sigma_{\delta'}}\} < \infty$ for every $\delta' \in (0,\delta)$;*

(ii) *There exist $\delta \in (0,\frac{\pi}{2})$ and $C > 0$ such that $\Sigma_{\frac{\pi}{2}+\delta} \subseteq \rho(A)$ and*

$$\left\|(\lambda I_X - A)^{-1}\right\|_{\mathcal{L}(X)} \leq \frac{C}{|\lambda|}, \quad \lambda \in \Sigma_{\frac{\pi}{2}+\delta}.$$

The theorem above is proved in [Paz83, page 61, Theorem 5.2(a),(c)]. The next result gives an efficient way of checking the conditions in Theorem 1.4.11. We remark that part of the proof can be found in [CHA+87, page 123, Lemma 5.1].

Proposition 1.4.12 *Let $\{S(t)\}_{t\geq 0}$ be a uniformly bounded, strongly continuous semigroup on X with infinitesimal generator A. If $\rho(A)$ contains the imaginary axis including 0 and if there exists $C>0$ such that*

$$\left\|(\mathrm{i}\omega I_X - A)^{-1}\right\|_{\mathcal{L}(X)} \leq \frac{C}{|\omega|}, \quad \omega \in \mathbb{R}\backslash\{0\}, \tag{1.4}$$

then there exists $\delta_0 \in (0, \frac{\pi}{2})$ such that $\Sigma_{\frac{\pi}{2}+\delta_0} \subseteq \rho(A)$. Moreover, for every $\delta \in (0, \frac{\pi}{2}+\delta_0)$ there exists $C_\delta > 0$ such that

$$\left\|(\lambda I_X - A)^{-1}\right\|_{\mathcal{L}(X)} \leq \frac{C_\delta}{|\lambda|}, \quad \lambda \in \overline{\Sigma}_\delta.$$

In particular, A is the infinitesimal generator of an analytic semigroup.

PROOF: Firstly, we note that $\rho(A)$ contains $\{z \in \mathbb{C};\, \mathrm{Re} z > 0\}$ and that there exists $M \geq 1$ such that

$$\left\|(zI_X - A)^{-1}\right\|_{\mathcal{L}(X)} \leq \frac{M}{\mathrm{Re} z}, \quad \mathrm{Re} z > 0, \tag{1.5}$$

see [Paz83, page 20, Remark 5.4]. Secondly, we show that there exists $\delta_0 \in (0, \frac{\pi}{2})$ such that

$$\left\{z \in \mathbb{C}\backslash\{0\};\, \frac{\pi}{2} - \delta_0 < |\mathrm{Arg}(z)| < \frac{\pi}{2} + \delta_0\right\} \subseteq \rho(A). \tag{1.6}$$

Let $\omega \in \mathbb{R}\backslash\{0\}$. If $\lambda \in \mathbb{C}$ is such that $|\lambda| < \frac{1}{C}|\omega|$, then $\|\lambda(\mathrm{i}\omega I_X - A)^{-1}\| < 1$ by (1.4) and hence, $I_X + \lambda(\mathrm{i}\omega I_X - A)^{-1} : X \to X$ is invertible with

$$\left(I_X + \lambda(\mathrm{i}\omega I_X - A)^{-1}\right)^{-1} = \sum_{k=0}^{\infty} (-\lambda)^k \left((\mathrm{i}\omega I_X - A)^{-1}\right)^k, \tag{1.7}$$

where the series is convergent in $\mathcal{L}(X)$, see [Kre78, page 375, Theorem 7.3-1]. In this case $\lambda + \mathrm{i}\omega \in \rho(A)$, since

$$(\lambda + \mathrm{i}\omega)I_X - A = \left(I_X + \lambda(\mathrm{i}\omega I_X - A)^{-1}\right)(\mathrm{i}\omega I_X - A). \tag{1.8}$$

Moreover, $\lambda + \mathrm{i}\omega \in B(\mathrm{i}\omega; \frac{1}{C}|\omega|)$. It follows that (1.6) holds with $\delta_0 := \arctan \frac{1}{C}$. At this point we have proved that $\Sigma_{\frac{\pi}{2}+\delta_0} \subseteq \rho(A)$.
Thirdly, using (1.5) we observe that for every $\delta \in (0, \frac{\pi}{2})$ there exists $M_\delta > 0$, namely $M_\delta := \frac{M}{\cos\delta}$, such that for every $z \in \overline{\Sigma}_\delta\backslash\{0\}$,

$$\left\|(zI_X - A)^{-1}\right\|_{\mathcal{L}(X)} \leq \frac{M}{\mathrm{Re} z} = \frac{M}{|z|\cos(\mathrm{Arg}(z))} \leq \frac{M}{|z|\cos\delta} = \frac{M_\delta}{|z|}. \tag{1.9}$$

Finally, we show that for every $\delta' \in (0, \delta_0)$ there exists $K_{\delta'} > 0$ such that for every $z \in \mathbb{C}\backslash\{0\}$ with $|\operatorname{Arg}(z)| \in [\frac{\pi}{2} - \delta', \frac{\pi}{2} + \delta']$,

$$\left\|(zI_X - A)^{-1}\right\|_{\mathcal{L}(X)} \leq \frac{K_{\delta'}}{|z|}. \tag{1.10}$$

Let $\delta' \in (0, \delta_0)$ and $z \in \mathbb{C}\backslash\{0\}$ with $|\operatorname{Arg}(z)| \in [\frac{\pi}{2} - \delta', \frac{\pi}{2} + \delta']$. Let $\lambda := \operatorname{Re} z$ and $\omega := \operatorname{Im} z$. Then we have

$$\frac{|\lambda|}{|\omega|} \leq \tan \delta' < \tan \delta_0 = \frac{1}{C}. \tag{1.11}$$

Therefore we can use (1.8), (1.7), (1.4), and (1.11) to obtain

$$\begin{aligned}\left\|(zI_X - A)^{-1}\right\|_{\mathcal{L}(X)} &\leq \left\|(\mathrm{i}\omega I_X - A)^{-1}\right\|_{\mathcal{L}(X)} \cdot \left(1 - |\lambda| \left\|(\mathrm{i}\omega I_X - A)^{-1}\right\|_{\mathcal{L}(X)}\right)^{-1} \\ &\leq \frac{C}{|\omega|} \cdot \frac{1}{1 - C \tan \delta'} \leq \frac{C}{|z| \cos \delta'} \cdot \frac{1}{1 - C \tan \delta'}.\end{aligned}$$

This shows that (1.10) holds with $K_{\delta'} := \frac{C}{\cos \delta'(1 - C \tan \delta')}$. Now the proof is finished by combining (1.9) and (1.10), and noting that we are in the situation of Theorem 1.4.11(ii). □

For the following perturbation result see [Paz83, page 81, Corollary 2.2].

Proposition 1.4.13 *Let $A : \mathrm{D}(A) \subseteq X \to X$ be the infinitesimal generator of an analytic semigroup on X. If B belongs to $\mathcal{L}(X)$, then $A+B$ is the infinitesimal generator of an analytic semigroup on X.*

Definition 1.4.14 Let $A : \mathrm{D}(A) \subseteq X \to X$ be the infinitesimal generator of an analytic semigroup on X. The space $(X, \mathrm{D}(A))_{\alpha,p}$ with $\alpha \in (0, 1)$ and $p \in [1, \infty)$ denotes the *real interpolation space between X and* $\mathrm{D}(A)$, and is defined by

$$(X, \mathrm{D}(A))_{\alpha,p} := \left\{x \in X;\ \int_c^\infty t^{p\alpha - 1} \left\|A(tI_X - A)^{-1} x\right\|_X^p \,\mathrm{d}t < \infty\right\},$$

$$\|x\|_{(X,\mathrm{D}(A))_{\alpha,p}} := \|x\|_X + \left(\int_c^\infty t^{p\alpha - 1} \|A(tI_X - A)^{-1} x\|_X^p \,\mathrm{d}t\right)^{\frac{1}{p}},$$

where $c > 1$ is sufficiently large.

The norm $\|\cdot\|_{(X,\mathrm{D}(A))_{\alpha,p}}$ is equivalent to another norm on $(X, \mathrm{D}(A))_{\alpha,p}$ that does not depend on c, see [Lun95, page 49, Proposition 2.2.6]. We remark that $(X, \mathrm{D}(A))_{\alpha,p}$ is a Banach space, see [Lun95, pages 18, 46, Propositions 1.2.4, 2.2.2]. For the following three lemmas we refer to [Lun95, page 47, Corollary 2.2.3(i),(ii),(iv)].

Lemma 1.4.15 *Let $A : \mathrm{D}(A) \subseteq X \to X$ and $B : \mathrm{D}(B) \subseteq X \to X$ be infinitesimal generators of analytic semigroups on X with $\mathrm{D}(A) = \mathrm{D}(B)$ and such that the graph norms of A and B are equivalent. Then*

$$(X, \mathrm{D}(A))_{\alpha,p} = (X, \mathrm{D}(B))_{\alpha,p}, \quad \alpha \in (0,1),\ p \in [1,\infty),$$

and the corresponding norms are equivalent.

Lemma 1.4.16 *Let $A : \mathrm{D}(A) \subseteq X \to X$ be the infinitesimal generator of an analytic semigroup on X. For all $\alpha, \gamma \in (0,1)$ and $p_1, p_2 \in [1,\infty)$ such that $\alpha < \gamma$ and $p_1 < p_2$, it holds that*

$$\mathrm{D}(A) \subseteq (X, \mathrm{D}(A))_{\gamma,p_1} \subseteq (X, \mathrm{D}(A))_{\gamma,p_2} \subseteq (X, \mathrm{D}(A))_{\alpha,p_1} \subseteq X,$$

where the inclusions are continuous embeddings and where $\mathrm{D}(A)$ is endowed with the graph norm of A.

Lemma 1.4.17 *Let $A : \mathrm{D}(A) \subseteq X \to X$ be the infinitesimal generator of an analytic semigroup on X. Then $\mathrm{D}(A)$ is dense in $(X, \mathrm{D}(A))_{\alpha,p}$ for all $\alpha \in (0,1)$ and $p \in [1,\infty)$.*

The next proposition states an important interpolation property, see [Lun95, page 19, Proposition 1.2.6].

Proposition 1.4.18 *Let X_i be a Banach space and $A_i : \mathrm{D}(A_i) \subseteq X_i \to X_i$ the infinitesimal generator of an analytic semigroup on X_i for $i = 1, 2$. If $T : X_1 \to X_2$ is a bounded linear operator such that $T : (\mathrm{D}(A_1), \|\cdot\|_{A_1}) \to (\mathrm{D}(A_2), \|\cdot\|_{A_2})$ is also bounded, then $T : (X_1, \mathrm{D}(A_1))_{\alpha,p} \to (X_2, \mathrm{D}(A_2))_{\alpha,p}$ is bounded for all $\alpha \in (0,1)$ and $p \in [1,\infty)$ and*

$$\|T\|_{\mathcal{L}((X_1,\mathrm{D}(A_1))_{\alpha,p},(X_2,\mathrm{D}(A_2))_{\alpha,p})} \leq \big(\|T\|_{\mathcal{L}(X_1,X_2)}\big)^{1-\alpha} \big(\|T\|_{\mathcal{L}(\mathrm{D}(A_1),\mathrm{D}(A_2))}\big)^{\alpha}.$$

For the following result we refer to [Ama95, page 289, Theorem 2.1.3].

Lemma 1.4.19 *Let $\{S(t)\}_{t\geq 0}$ be a strongly continuous semigroup on X with infinitesimal generator A. Let $\alpha \in (0,1)$ and $p \in [1,\infty)$. Let $A_{\alpha,p} : \mathrm{D}(A_{\alpha,p}) \subseteq (X, \mathrm{D}(A))_{\alpha,p} \to (X, \mathrm{D}(A))_{\alpha,p}$ be defined by*

$$\mathrm{D}(A_{\alpha,p}) := \big\{x \in \mathrm{D}(A);\ Ax \in (X, \mathrm{D}(A))_{\alpha,p}\big\},$$

$$A_{\alpha,p}x := Ax, \quad x \in \mathrm{D}(A_{\alpha,p}).$$

Then $\{S(t)|_{(X,\mathrm{D}(A))_{\alpha,p}}\}_{t\geq 0}$ is a strongly continuous semigroup on $(X, \mathrm{D}(A))_{\alpha,p}$ with infinitesimal generator $A_{\alpha,p}$. Moreover, $\rho(A) \subseteq \rho(A_{\alpha,p})$ and

$$\big(\lambda I_{(X,\mathrm{D}(A))_{\alpha,p}} - A_{\alpha,p}\big)^{-1} = (\lambda I_X - A)^{-1}|_{(X,\mathrm{D}(A))_{\alpha,p}}, \quad \lambda \in \rho(A).$$

We remark that $A_{\alpha,p}$ is called *the part of* A *in* $(X, \mathrm{D}(A))_{\alpha,p}$.

Proposition 1.4.20 *Let* $A : \mathrm{D}(A) \subseteq X \to X$ *be the infinitesimal generator of an analytic semigroup on* X. *Then*

$$C^{0,\gamma}\big([0,T]; \mathrm{D}(A)\big) \cap C^{1,\gamma}\big([0,T]; X\big)$$

is a subset of

$$C^{[\gamma+1-\alpha],\gamma+1-\alpha-[\gamma+1-\alpha]}\big([0,T]; (X, \mathrm{D}(A))_{\alpha,2}\big),$$

for all $T > 0$, $p \in [1, \infty)$, *and* $\alpha, \gamma \in (0,1)$ *such that* $\alpha \neq \gamma$.

The proposition above is proved in [Lun95, page 14, Proposition 1.1.5]. For the next result we refer to [Lun95, page 28, Remark 1.2.16].

Theorem 1.4.21 (Reiteration Theorem) *Let* $A : \mathrm{D}(A) \subseteq X \to X$ *be the infinitesimal generator of an analytic semigroup on* X. *Then for all* $\alpha, \gamma_1, \gamma_2 \in (0,1)$ *and* $p \in [1, \infty)$, *it holds that*

$$\begin{aligned}
\big((X, \mathrm{D}(A))_{\gamma_1,p}, (X, \mathrm{D}(A))_{\gamma_2,p}\big)_{\alpha,p} &= (X, \mathrm{D}(A))_{(1-\alpha)\gamma_1+\alpha\gamma_2,p},\\
\big((X, \mathrm{D}(A))_{\gamma_1,p}, \mathrm{D}(A)\big)_{\alpha,p} &= (X, \mathrm{D}(A))_{(1-\alpha)\gamma_1+\alpha,p},\\
\big(X, (X, \mathrm{D}(A))_{\gamma_2,p}\big)_{\alpha,p} &= (X, \mathrm{D}(A))_{\alpha\gamma_2,p}.
\end{aligned}$$

We state the following proposition for its useful corollary, which is an immediate application of the reiteration theorem. We refer to [Lun91, pages 97, 114, 120, Theorems 4.2.6, 4.3.5, Corollary 4.3.12], [MCSA01, pages 123, 192, 250, Theorem 5.4.3, Proposition 8.1.1, Theorem 10.1.6], and [Ama95, page 266, Lemma 1.3.7].

Proposition 1.4.22 *Let* H *be a Hilbert space and let* $A : \mathrm{D}(A) \subseteq H \to H$ *be m-dissipative with* $0 \in \rho(A)$. *Then*

$$H = (H_{-1}, \mathrm{D}(A))_{\frac{1}{2},2},$$

and the corresponding norms are equivalent.

Corollary 1.4.23 *Let* H *be a Hilbert space and let* $A : \mathrm{D}(A) \subseteq H \to H$ *be m-dissipative with* $0 \in \rho(A)$. *Let* $\alpha \in (0,1)$. *Then*

$$\begin{aligned}
(H_{-1}, H)_{\alpha,2} &= (H_{-1}, \mathrm{D}(A))_{\frac{1}{2}\alpha,2},\\
(H, \mathrm{D}(A))_{\alpha,2} &= (H_{-1}, \mathrm{D}(A))_{\frac{1}{2}(\alpha+1),2},
\end{aligned}$$

and in both equalities the corresponding norms are equivalent.

Definition 1.4.24 Let $A : \mathrm{D}(A) \subseteq X \to X$ be the infinitesimal generator of an analytic semigroup on X that is uniformly bounded in a closed sector $\overline{\Sigma}_{\delta'}$ for some $\delta' \in (0,\pi)$, and let $0 \in \rho(A)$. The fractional power $(-A)^{-\alpha} : X \to X$ with $\alpha \in (0,1)$ is defined by

$$(-A)^{-\alpha}x := \frac{\sin(\pi\alpha)}{\pi} \int_0^\infty t^{-\alpha}(tI_X - A)^{-1}x \,\mathrm{d}t, \quad x \in X.$$

We remark that $(-A)^{-\alpha}$ is a bounded linear operator on X, see [Paz83, page 69]. Furthermore, $(-A)^{-\alpha}$ is one-to-one. For a proof we refer to [Paz83, page 72, Lemma 6.6].

Definition 1.4.25 Let $A : \mathrm{D}(A) \subseteq X \to X$ be the infinitesimal generator of an analytic semigroup on X that is uniformly bounded in a closed sector $\overline{\Sigma}_{\delta'}$ for some $\delta' \in (0,\pi)$, and let $0 \in \rho(A)$. The fractional power $(-A)^{\alpha} : \mathrm{D}((-A)^{\alpha}) \to X$ with $\alpha \in (0,1)$ is defined by

$$\mathrm{D}((-A)^{\alpha}) := \mathrm{Ran}((-A)^{-\alpha}),$$

$$(-A)^{\alpha}x := (-A)^{-\alpha}x, \quad x \in \mathrm{D}((-A)^{\alpha}).$$

For the sake of completeness we state $(-A)^1 := -A$, $(-A)^0 := I_X$, and $(-A)^{-1} := -A^{-1}$.

Proposition 1.4.26 *Let $A : \mathrm{D}(A) \subseteq X \to X$ be the infinitesimal generator of an analytic semigroup on X that is uniformly bounded in a closed sector $\overline{\Sigma}_{\delta'}$ for some $\delta' \in (0,\pi)$, and let $0 \in \rho(A)$. Then the following holds:*

(i) $(-A)^{-\alpha}(-A)^{\alpha} = I_{\mathrm{D}((-A)^{\alpha})}$ *for every* $\alpha \in (0,1)$*;*

(ii) $(-A)^{\alpha-1}(-A)x = (-A)^{\alpha}x$ *for all* $\alpha \in (0,1)$ *and* $x \in \mathrm{D}(A)$*.*

A proof of the proposition above can be found in [Paz83, page 72, Theorem 6.8(d)]. For a proof of the next theorem we refer to [Lun91, pages 114, 97, 120, Theorems 4.3.5, 4.2.6, Corollary 4.3.12].

Theorem 1.4.27 *Let $(H, \langle\cdot,\cdot\rangle_H, \|\cdot\|_H)$ be a Hilbert space and $A : \mathrm{D}(A) \subseteq H \to H$ the infinitesimal generator of an analytic semigroup on H that is uniformly bounded in a closed sector $\overline{\Sigma}_{\delta'}$ for some $\delta' \in (0,\pi)$, and let $0 \in \rho(A)$. Then*

$$(H, \mathrm{D}(A))_{\alpha,2} = \mathrm{D}((-A)^{\alpha}), \quad \alpha \in (0,1).$$

Moreover, the norm of $(H, \mathrm{D}(A))_{\alpha,2}$ is equivalent to the norm $x \mapsto \|(-A)^{\alpha}x\|_H$ on $\mathrm{D}((-A)^{\alpha})$.

Lemma 1.4.28 *Let $(H, \langle\cdot,\cdot\rangle_H, \|\cdot\|_H)$ be a Hilbert space and $A : \mathrm{D}(A) \subseteq H \to H$ the infinitesimal generator of an analytic semigroup on H that is uniformly bounded in a closed sector $\overline{\Sigma}_{\delta'}$ for some $\delta' \in (0,\pi)$, and let $0 \in \rho(A)$. Then*

$$((-A)^\alpha)^* = (-A^*)^\alpha, \quad \alpha \in (-1,1).$$

For the lemma above we refer to [Lun91, page 100]. The next proposition is proved in [Paz83, page 74, Theorem 6.13(a)-(c)]. A proof of the succeeding corollary can be found in [Lun95, page 51, Proposition 2.2.9].

Proposition 1.4.29 *Let $A : \mathrm{D}(A) \subseteq X \to X$ be the infinitesimal generator of an analytic semigroup on X that is uniformly bounded in a closed sector $\overline{\Sigma}_{\delta'}$ for some $\delta' \in (0,\pi)$, and let $0 \in \rho(A)$. Then the following holds:*

(i) *$S(t)x \in \mathrm{D}(A)$ for all $t > 0$ and $x \in X$;*

(ii) *$(-A)^\alpha S(t)x = S(t)(-A)^\alpha x$ for all $t \geq 0$, $\alpha \in [0,1]$, and $x \in \mathrm{D}((-A)^\alpha)$;*

(iii) *There exists $\delta > 0$ with the property that for every $\alpha \in [0,1]$ there exists $M_\alpha > 0$ such that $\|(-A)^\alpha S(t)\|_{\mathcal{L}(X)} \leq M_\alpha t^{-\alpha} \mathrm{e}^{-\delta t}$ for every $t > 0$.*

Corollary 1.4.30 *Let $A : \mathrm{D}(A) \subseteq X \to X$ be the infinitesimal generator of an analytic semigroup on X that is uniformly bounded in a closed sector $\overline{\Sigma}_{\delta'}$ for some $\delta' \in (0,\pi)$, and let $0 \in \rho(A)$. Then there exists $\delta > 0$ with the property that for all $\alpha, \gamma \in (0,1)$ with $\alpha \leq \gamma$ and $p \in [1,\infty)$ there exist $M_{\alpha,\gamma,p} > 0$ and $M'_{\alpha,\gamma,p} > 0$ such that the following holds:*

(i) *It holds that*

$$\|S(t)\|_{\mathcal{L}((X,\mathrm{D}(A))_{\alpha,p},(X,\mathrm{D}(A))_{\gamma,p})} \leq M_{\alpha,\gamma,p} t^{\alpha-\gamma} \mathrm{e}^{-\delta t}, \quad t > 0;$$

(ii) *It holds that*

$$\|A^n S(t)\|_{\mathcal{L}((X,\mathrm{D}(A))_{\gamma,p},(X,\mathrm{D}(A))_{\alpha,p})} \leq M'_{\alpha,\gamma,p} t^{-n+\gamma-\alpha} \mathrm{e}^{-\delta t}, \quad t > 0,\ n = 1,2.$$

Proposition 1.4.31 *Let $\{S(t)\}_{t\geq 0}$ be an analytic semigroup on X that is uniformly bounded in a closed sector $\overline{\Sigma}_{\delta'}$ for some $\delta' \in (0,\pi)$. Let the infinitesimal generator A be such that $0 \in \rho(A)$. Let $\alpha \in (0,1)$ and $p, r \in [1,\infty)$. Let $f : [0,\infty) \to X$ belong to $L^p(0,T;(X,\mathrm{D}(A))_{\alpha,r})$ for every $T > 0$. Let $T > 0$. Then the following holds:*

(i) *If x belongs to $(X,\mathrm{D}(A))_{\alpha,r}$, then the mapping $t \mapsto S(t)x$ belongs to $C^{0,\alpha}([0,T];X)$;*

(ii) *If* x *belongs to* $\mathrm{D}(A)$, *then the mapping* $t \mapsto S(t)x$ *belongs to* $C^{0,1-\alpha}\big([0,T];(X,\mathrm{D}(A))_{\alpha,r}\big)$;

(iii) *Let* $\gamma \in (0,\alpha]$. *If* x *belongs to* $(X,\mathrm{D}(A))_{\alpha,r}$, *then the mapping* $t \mapsto S(t)x$ *belongs to* $C^{0,\alpha-\gamma}\big([0,T];(X,\mathrm{D}(A))_{\gamma,r}\big)$;

(iv) *Let* $\gamma \in (0,\alpha)$. *If* $p = 1$, *then the mapping* $t \mapsto \int_0^t AS(t-s)f(s)\,\mathrm{d}s$ *belongs to* $L^q\big(0,T;(X,\mathrm{D}(A))_{\gamma,r}\big)$ *for every* $q \in \big[1,\frac{1}{1-(\alpha-\gamma)}\big)$;

(v) *Let* $\gamma \in (0,\alpha)$. *If* $p \in \big(1,\frac{1}{\alpha-\gamma}\big)$, *then the mapping* $t \mapsto \int_0^t AS(t-s)f(s)\,\mathrm{d}s$ *belongs to* $L^q\big(0,T;(X,\mathrm{D}(A))_{\gamma,r}\big)$ *where* $q := \frac{p}{1-(\alpha-\gamma)p}$;

(vi) *Let* $\gamma \in (0,\alpha)$. *If* $p = \frac{1}{\alpha-\gamma}$, *then the mapping* $t \mapsto \int_0^t AS(t-s)f(s)\,\mathrm{d}s$ *belongs to* $L^q\big(0,T;(X,\mathrm{D}(A))_{\gamma,r}\big)$ *for every* $q \in [1,\infty)$;

(vii) *Let* $\gamma \in (0,\alpha)$. *If* $p \in \big(\frac{1}{\alpha-\gamma},\infty\big)$, *then the mapping* $t \mapsto \int_0^t AS(t-s)f(s)\,\mathrm{d}s$ *belongs to* $C_0^{0,\alpha-\gamma-\frac{1}{p}}\big([0,T];(X,\mathrm{D}(A))_{\gamma,r}\big)$.

PROOF: For property (i) we refer to [Lun95, pages 47, 48, Corollary 2.2.3, Remark 2.2.5]. Property (ii) follows from [Lun95, pages 13, 19, Proposition 1.1.4(i), Corollary 1.2.7], since the mapping $t \mapsto S(t)x$ belongs, for every $x \in \mathrm{D}(A)$, to

$$C^1\big([0,T];X\big) \cap C\big([0,T];(\mathrm{D}(A),\|\cdot\|_A)\big).$$

For property (iii) we observe that if x belongs to $(X,\mathrm{D}(A))_{\alpha,r}$, then using property (i) we have that the mapping $t \mapsto S(t)$ belongs to

$$C^{0,\alpha}\big([0,T];X\big) \cap C\big([0,T];(X,\mathrm{D}(A))_{\alpha,r}\big).$$

Now [Lun95, pages 13, Proposition 1.1.4(ii)] implies that the mapping $t \mapsto S(t)$ belongs to

$$C^{0,\alpha(1-\frac{\gamma}{\alpha})}\Big([0,T];\big(X,(X,\mathrm{D}(A))_{\alpha,r}\big)_{\frac{\gamma}{\alpha},r}\Big).$$

By Theorem 1.4.21 we have

$$\big(X,(X,\mathrm{D}(A))_{\alpha,r}\big)_{\frac{\gamma}{\alpha},r} = (X,\mathrm{D}(A))_{\gamma,r},$$

and property (iii) follows in case $\gamma < \alpha$. The case $\alpha = \gamma$ is obvious.
For the proof of properties (iv)-(vii) let $\gamma \in (0,\alpha)$. To prove property (iv) let $q \in \big[1,\frac{1}{1-(\alpha-\gamma)}\big)$. We use Corollary 1.4.30(ii) and Proposition 1.3.4(i) with α

replaced by $\alpha-\gamma$ and f replaced by the mapping $t \mapsto \|f(t)\|_{(X,\mathrm{D}(A))_{\alpha,r}}$ to obtain that there exists $M>0$ such that

$$\begin{aligned}
&\int_0^T \left\| \int_0^t AS(t-s)f(s)\,\mathrm{d}s \right\|^q_{(X,\mathrm{D}(A))_{\gamma,r}} \mathrm{d}t \\
&\leq \int_0^T \left(\int_0^t \|AS(t-s)\|_{\mathcal{L}((X,\mathrm{D}(A))_{\alpha,r},(X,\mathrm{D}(A))_{\gamma,r})} \|f(s)\|_{(X,\mathrm{D}(A))_{\alpha,r}}\,\mathrm{d}s \right)^q \mathrm{d}t \\
&\leq \int_0^T \left(\int_0^t M(t-s)^{-1+\alpha-\gamma} \|f(s)\|_{(X,\mathrm{D}(A))_{\alpha,r}}\,\mathrm{d}s \right)^q \mathrm{d}t \\
&= M^q\Gamma(\alpha-\gamma) \int_0^T \left(\int_0^t g_{\alpha-\gamma}(t-s) \|f(s)\|_{(X,\mathrm{D}(A))_{\alpha,r}}\,\mathrm{d}s \right)^q \mathrm{d}t \\
&\leq M^q\Gamma(\alpha-\gamma)c^q \|f\|^q_{L^p([0,T];(X,\mathrm{D}(A))_{\alpha,r})} < \infty.
\end{aligned}$$

Properties (v) and (vi) are proved analogously using Proposition 1.3.4(ii) respectively (iii).
For property (vii) we first observe that $\int_0^t AS(t-s)f(s)\,\mathrm{d}s$ exists as a Bochner integral in $(X,\mathrm{D}(A))_{\gamma,r}$ for every $t>0$. Indeed, if $p'>1$ is such that $\frac{1}{p}+\frac{1}{p'}=1$, then using Corollary 1.4.30(ii), the Hölder inequality, and the fact that $(\alpha-\gamma-1)p'>-1$, we have that there exists $M>0$ such that

$$\begin{aligned}
&\int_0^t \|AS(t-s)f(s)\|_{(X,\mathrm{D}(A))_{\gamma,r}}\,\mathrm{d}s \\
&\leq \int_0^t M(t-s)^{-1+\alpha-\gamma}\|f(s)\|_{(X,\mathrm{D}(A))_{\alpha,r}}\,\mathrm{d}s \\
&\leq M\|f\|_{L^p([0,T];(X,\mathrm{D}(A))_{\alpha,r})} \left(\int_0^t s^{(\alpha-\gamma-1)p'}\,\mathrm{d}s \right)^{\frac{1}{p'}} < \infty, \qquad t>0.
\end{aligned}$$

Furthermore, we observe that

$$\int_0^\infty \left(\tau^{\alpha-\gamma-1} - (1+\tau)^{\alpha-\gamma-1}\right)^{p'} \mathrm{d}\tau < \infty. \tag{1.12}$$

Indeed, since $(\alpha-\gamma-1)p'>-1$, we have

$$0 \leq \int_0^1 \left(\tau^{\alpha-\gamma-1} - (1+\tau)^{\alpha-\gamma-1}\right)^{p'} \mathrm{d}\tau \leq \int_0^1 \tau^{(\alpha-\gamma-1)p'}\,\mathrm{d}\tau < \infty,$$

and, using the mean value theorem and the fact that $(\alpha-\gamma-2)p'<-1$,

$$\int_1^\infty \left(\tau^{\alpha-\gamma-1} - (1+\tau)^{\alpha-\gamma-1}\right)^{p'} \mathrm{d}\tau \leq (1-(\alpha-\gamma)) \int_1^\infty \tau^{(\alpha-\gamma-2)p'}\,\mathrm{d}\tau < \infty.$$

Using Corollary 1.4.30(ii), the Hölder inequality, the substitution $\tau := \frac{s-\sigma}{t-s}$, and (1.12), it follows that there exist $M > 0$ and $c_1, c_2 > 0$ such that for all $0 < s < t \leq T$,

$$\begin{aligned}
&\left\| \int_0^t AS(t-\sigma)f(\sigma)\,\mathrm{d}\sigma - \int_0^s AS(s-\sigma)f(\sigma)\,\mathrm{d}\sigma \right\|_{(X,\mathrm{D}(A))_{\gamma,r}} \\
&= \left\| \int_0^s A\big(S(t-\sigma) - S(s-\sigma)\big)f(\sigma)\,\mathrm{d}\sigma + \int_s^t AS(t-\sigma)f(\sigma)\,\mathrm{d}\sigma \right\|_{\mathrm{D}_A(\gamma,r)} \\
&= \left\| \int_0^s \left(\int_{s-\sigma}^{t-\sigma} A^2 S(\tau) f(\sigma)\,\mathrm{d}\tau \right) \mathrm{d}\sigma + \int_s^t AS(t-\sigma)f(\sigma)\,\mathrm{d}\sigma \right\|_{\mathrm{D}_A(\gamma,r)} \\
&\leq \int_0^s \left(\int_{s-\sigma}^{t-\sigma} \|A^2S(\tau)\|_{\mathcal{L}((X,\mathrm{D}(A))_{\alpha,r},(X,\mathrm{D}(A))_{\gamma,r})}\,\mathrm{d}\tau \right) \|f(\sigma)\|_{(X,\mathrm{D}(A))_{\alpha,r}}\,\mathrm{d}\sigma + \\
&\quad \int_s^t \|AS(t-\sigma)\|_{\mathcal{L}((X,\mathrm{D}(A))_{\alpha,r},(X,\mathrm{D}(A))_{\gamma,r})} \|f(\sigma)\|_{(X,\mathrm{D}(A))_{\alpha,r}}\,\mathrm{d}\sigma \\
&\leq \int_0^s \left(\int_{s-\sigma}^{t-\sigma} M\tau^{-2+\alpha-\gamma}\,\mathrm{d}\tau \right) \|f(\sigma)\|_{(X,\mathrm{D}(A))_{\alpha,r}}\mathrm{d}\sigma + \\
&\quad \int_s^t M(t-\sigma)^{-1+\alpha-\gamma} \|f(\sigma)\|_{(X,\mathrm{D}(A))_{\alpha,r}}\,\mathrm{d}\sigma \\
&= \frac{M}{1-(\alpha-\gamma)} \int_0^s \left((s-\sigma)^{\alpha-\gamma-1} - (t-\sigma)^{\alpha-\gamma-1} \right) \|f(\sigma)\|_{(X,\mathrm{D}(A))_{\alpha,r}}\,\mathrm{d}\sigma + \\
&\quad M \int_s^t (t-\sigma)^{\alpha-\gamma-1} \|f(\sigma)\|_{(X,\mathrm{D}(A))_{\alpha,r}}\,\mathrm{d}\sigma \\
&\leq \frac{M}{1-(\alpha-\gamma)} \left(\left(\int_0^s \left((s-\sigma)^{\alpha-\gamma-1} - (t-\sigma)^{\alpha-\gamma-1} \right)^{p'} \mathrm{d}\sigma \right)^{\frac{1}{p'}} + \right. \\
&\quad \left. \left(\int_s^t (t-\sigma)^{(\alpha-\gamma-1)p'}\,\mathrm{d}\sigma \right)^{\frac{1}{p'}} \right) \|f\|_{L^p([0,T];(X,\mathrm{D}(A))_{\alpha,r})} \\
&\leq \frac{M}{1-(\alpha-\gamma)} \left((t-s)^{\alpha-\gamma-\frac{1}{p}} \left(\int_0^{\frac{s}{t-s}} \left(\tau^{\alpha-\gamma-1} - (1+\tau)^{\alpha-\gamma-1} \right)^{p'} \mathrm{d}\tau \right)^{\frac{1}{p'}} + \right. \\
&\quad \left. c_1 (t-s)^{\alpha-\gamma-\frac{1}{p}} \right) \|f\|_{L^p([0,T];(X,\mathrm{D}(A))_{\alpha,r})} \\
&\leq \frac{M}{1-(\alpha-\gamma)} \left(\left(\int_0^\infty \left(\tau^{\alpha-\gamma-1} - (1+\tau)^{\alpha-\gamma-1} \right)^{p'} \mathrm{d}\tau \right)^{\frac{1}{p'}} + c_1 \right) \cdot \\
&\quad \|f\|_{L^p([0,T];(X,\mathrm{D}(A))_{\alpha,r})} (t-s)^{\alpha-\gamma-\frac{1}{p}} \\
&= \frac{M}{1-(\alpha-\gamma)} (c_2 + c_1) \|f\|_{L^p([0,T];(X,\mathrm{D}(A))_{\alpha,r})} (t-s)^{\alpha-\gamma-\frac{1}{p}}.
\end{aligned}$$

This shows that the mapping $t \mapsto \int_0^t AS(s-t)f(s)\,\mathrm{d}s$ belongs to

$$C^{0,\alpha-\gamma-\frac{1}{p}}\big((0,T];(X,\mathrm{D}(A))_{\gamma,r}\big).$$

Moreover, using Corollary 1.4.30(ii) we have that there exists $M>0$ such that

$$\begin{aligned}
&\left\|\int_0^t AS(t-s)f(s)\,\mathrm{d}s\right\|_{(X,\mathrm{D}(A))_{\gamma,r}}\\
&\leq \int_0^t \big\|AS(t-s)f(s)\big\|_{(X,\mathrm{D}(A))_{\gamma,r}}\,\mathrm{d}s\\
&\leq M\int_0^t (t-s)^{-1+\alpha-\gamma}\|f(s)\|_{(X,\mathrm{D}(A))_{\alpha,r}}\,\mathrm{d}s\\
&= M\Gamma(\alpha-\gamma)\int_0^t g_{\alpha-\gamma}(t-s)\|f(s)\|_{(X,\mathrm{D}(A))_{\alpha,r}}\,\mathrm{d}s, \quad t\in(0,T].
\end{aligned}$$

As the mapping $t \mapsto \int_0^t g_{\alpha-\gamma}(t-s)\|f(s)\|_{(X,\mathrm{D}(A))_{\alpha,r}}\,\mathrm{d}s$ belongs to $C_0^{0,\alpha-\gamma-\frac{1}{p}}[0,T]$ as a result of Proposition 1.3.4(iv), it makes sense to define

$$\left.\int_0^t AS(t-s)f(s)\,\mathrm{d}s\right|_{t=0} := 0.$$

From this definition combined with the above it follows that there exists $c>0$ such that

$$\left\|\int_0^t AS(t-s)f(s)\,\mathrm{d}s\right\|_{(X,\mathrm{D}(A))_{\gamma,r}} \leq ct^{\alpha-\gamma-\frac{1}{p}}, \quad t\in(0,T].$$

Hence, the mapping $t \mapsto \int_0^t AS(t-s)f(s)\,\mathrm{d}s$ belongs to

$$C_0^{0,\alpha-\gamma-\frac{1}{p}}\big([0,T];(X,\mathrm{D}(A))_{\gamma,r}\big).$$

□

1.5 Trace class operators

Definition 1.5.1 A *trace class operator* on a separable Hilbert space $(H,\langle\cdot,\cdot\rangle,\|\cdot\|)$ is an operator $T\in\mathcal{L}(H)$ with the property that there exist two sequences $\{\varphi_n\}_{n=1}^\infty$ and $\{\psi_n\}_{n=1}^\infty$ in H such that $\sum_{n=1}^\infty \|\varphi_n\|\,\|\psi_n\|<\infty$ and

$$Tx = \sum_{n=1}^\infty \langle x,\psi_n\rangle\varphi_n, \quad x\in H.$$

Proposition 1.5.2 *If T is a trace class operator on a separable Hilbert space $(H, \langle\cdot,\cdot\rangle)$ and if $\{e_n\}_{n=1}^{\infty}$ is a complete orthonormal system in H, then the series $\sum_{n=1}^{\infty}\langle Te_n, e_n\rangle$ is absolutely convergent and independent of $\{e_n\}_{n=1}^{\infty}$.*

Definition 1.5.3 If T is a trace class operator on a separable Hilbert space $(H, \langle\cdot,\cdot\rangle)$ and if $\{e_n\}_{n=1}^{\infty}$ is a complete orthonormal system in H, then the *trace* of T is defined by

$$\operatorname{Tr} T := \sum_{n=1}^{\infty}\langle Te_n, e_n\rangle.$$

Proposition 1.5.4 *Let $(H, \langle\cdot,\cdot\rangle)$ be a separable Hilbert space. A nonnegative*[1] *operator $T \in \mathcal{L}(H)$ is trace class if and only if $\sum_{n=1}^{\infty}\langle Te_n, e_n\rangle < \infty$ for some complete orthonormal system $\{e_n\}_{n=1}^{\infty}$ in H.*

We refer to [DPZ92, pages 415-417, Appendix C, Propositions C.1, C.3].

[1]Nonnegative in the sense that $\langle Tx, x\rangle \geq 0$ for every $x \in H$.

Chapter 2

Pointwise versions of solutions to Cauchy problems in L^p-spaces

This chapter contains some technical results on existence and regularity of pointwise versions of solutions to Cauchy problems in L^p-spaces. These results are needed in later chapters to return from the abstract semigroup setting to the Volterra equation.

2.1 Introduction

Let $(X, \|\cdot\|)$ be a complex Banach space, $(\Omega, \mathcal{F}, \mu)$ a σ-finite measure space, and $p \in [1, \infty)$. In this chapter we distinguish between a function $\varphi : \Omega \to X$ and its equivalence class. The equivalence class of φ is denoted by $[\varphi]$ and consists of functions $\psi : \Omega \to X$ such that $\psi(\omega) = \varphi(\omega)$ for almost every $\omega \in \Omega$. Functions φ and ψ are called *versions* of the equivalence class $[\varphi]$. We also distinguish between L^p-spaces of functions and L^p-spaces of equivalence classes by the following notation:

$$\mathcal{L}^p(\Omega; X) := \left\{ \varphi : \Omega \to X;\ \varphi \text{ strongly measurable}, \int_\Omega \|\varphi(\omega)\|^p \,\mu(\mathrm{d}\omega) < \infty \right\},$$

$$L^p(\Omega; X) := \big\{ [\varphi];\ \varphi \in \mathcal{L}^p(\Omega; X) \big\}, \quad \big\|[\varphi]\big\|_{L^p(\Omega;X)} := \left(\int_\Omega \|\varphi(\omega)\|^p \,\mu(\mathrm{d}\omega) \right)^{\frac{1}{p}}.$$

In $\mathcal{L}^p(\Omega; X)$ we consider the Cauchy problem

$$\begin{aligned} \frac{\partial}{\partial t}\varphi(t,\omega) &= \big(\mathcal{A}\varphi(t,\cdot)\big)(\omega), \quad t > 0,\ \omega \in \Omega, \\ \varphi(0,\omega) &= \varphi_0(\omega), \qquad\qquad\qquad \omega \in \Omega, \end{aligned} \tag{2.1}$$

where $\mathcal{A} : \mathrm{D}(\mathcal{A}) \subseteq \mathcal{L}^p(\Omega; X) \to \mathcal{L}^p(\Omega; X)$ is a linear operator and φ_0 belongs to $\mathcal{L}^p(\Omega; X)$. Frequently, $\mathcal{A}$ is a partial differential operator with respect to ω and Ω

a domain in $\mathbb{R}^d$ for some $d \in \mathbb{N}$, but it need not be so. We assume that problem (2.1) can be rewritten in $L^p(\Omega; X)$ as the abstract Cauchy problem

$$\begin{aligned} \frac{\mathrm{d}}{\mathrm{d}t}[\varphi(t)] &= A[\varphi(t)], \quad t > 0, \\ [\varphi(0)] &= [\varphi_0], \end{aligned} \tag{2.2}$$

where $A : \mathrm{D}(A) \subseteq L^p(\Omega; X) \to L^p(\Omega; X)$ is the infinitesimal generator of a strongly continuous semigroup $\{S(t)\}_{t \geq 0}$ on $L^p(\Omega; X)$. Then problem (2.2) admits the mild solution $t \mapsto [\varphi(t)]$ given by $[\varphi(t)] = S(t)[\varphi_0]$ for every $t \geq 0$ and problem (2.1) seems to be solved. However, this is not the case, yet.
For every $t > 0$ problem (2.1) requires a version $\varphi(t, \cdot)$ of $[\varphi(t)]$. If $\mathcal{A}$ is a partial differential operator and $\{S(t)\}_{t \geq 0}$ exhibits some smoothing, then we simply choose $\varphi(t, \cdot)$ to be the unique version of $[\varphi(t)]$ such that the mapping $\omega \mapsto \varphi(t, \omega)$ is smooth for every $t > 0$. If $\mathcal{A}$ is not a partial differential operator, then the choice of a version $\varphi(t, \cdot)$ is less evident. To see that this is a nontrivial problem we consider an example in Section 2.2. More can be said if $\{S(t)\}_{t \geq 0}$ is an analytic semigroup. In Section 2.3 we work out the consequences of a result of E.M. Stein concerning versions of $[\varphi(t)]$ under analyticity assumptions. In Section 2.4 we return to the Cauchy problem (2.1).

2.2 An example

This section shows that the mild solution to problem (2.2) may not have a smooth version. We consider problem (2.2) in the space consisting of 1-periodic integrable functions, for which we introduce the following notation:

$$\begin{aligned} \mathcal{L}^1_1(\mathbb{R}) &:= \Big\{\varphi : \mathbb{R} \to \mathbb{R};\ \varphi \text{ Borel measurable, } \int_0^1 |\varphi(\omega)|\, \mathrm{d}\omega < \infty, \text{ and} \\ &\qquad \varphi(\omega + 1) = \varphi(\omega) \text{ for every } \omega \in \mathbb{R}\Big\}, \\ L^1_1(\mathbb{R}) &:= \{[\varphi];\ \varphi \in \mathcal{L}^1_1(\mathbb{R})\}, \qquad \big\|[\varphi]\big\|_{L^1_1(\mathbb{R})} := \int_0^1 |\varphi(\omega)|\, \mathrm{d}\omega. \end{aligned}$$

Example 2.2.1 Let $\mathcal{S}(t) : \mathcal{L}^1_1(\mathbb{R}) \to \mathcal{L}^1_1(\mathbb{R})$ and $S(t) : L^1_1(\mathbb{R}) \to L^1_1(\mathbb{R})$ for every $t \geq 0$ be given by respectively

$$\begin{aligned} \big(\mathcal{S}(t)\varphi\big)(\omega) &:= \varphi(t + \omega), \quad \omega \in \mathbb{R},\ \varphi \in \mathcal{L}^1_1(\mathbb{R}), \\ S(t)[\varphi] &:= [\mathcal{S}(t)\varphi], \qquad\qquad \varphi \in \mathcal{L}^1_1(\mathbb{R}). \end{aligned}$$

Then there exists a function φ_0 in $\mathcal{L}^1_1(\mathbb{R})$ with the property that if $\varphi : [0, \infty) \times \mathbb{R} \to \mathbb{R}$ is a Borel measurable function such that $\varphi(t, \cdot)$ is a version of $S(t)[\varphi_0]$ for every $t > 0$, then for almost every $\omega \in \mathbb{R}$, $\lim_{t \downarrow 0} \varphi(t, \omega)$ does not exist.

PROOF: To construct φ_0 let $\{\varphi_k\}_{k=1}^{\infty}$ be the sequence of functions in $\mathcal{L}_1^1(\mathbb{R})$ given by

$$\varphi_k(\omega) := \begin{cases} 1, & (\omega \mod 1) \in \cup_{j=0}^{2^k-1}\left[j\cdot 2^{-k}, j\cdot 2^{-k}+2^{-2k}\right], \\ 0, & \omega \text{ elsewhere in } \mathbb{R}. \end{cases}$$

Note that $\int_0^1 \varphi_k(\omega)\,\mathrm{d}\omega = 2^k\cdot 2^{-2k} = 2^{-k}$ for $k = 1, 2, \cdots$. Therefore Lebesgue's monotone convergence theorem implies that $\int_0^1 \sum_{k=1}^{\infty}\varphi_k(\omega)\,\mathrm{d}\omega < \infty$ and hence, $\sum_{k=1}^{\infty}\varphi_k(\omega)$ is a convergent series for almost every $\omega \in \mathbb{R}$. Now we define $\varphi_0 : \mathbb{R} \to \mathbb{R}$ by

$$\varphi_0(\omega) := \begin{cases} \sum_{k=1}^{\infty}\varphi_k(\omega), & \omega \in \mathbb{R} \text{ such that the series converges}, \\ 0, & \omega \text{ elsewhere in } \mathbb{R}. \end{cases}$$

It follows from the above that φ_0 belongs to $\mathcal{L}_1^1(\mathbb{R})$. Moreover, the series $\sum_{k=1}^{\infty}[\varphi_k]$ converges in $L_1^1(\mathbb{R})$, so that $[\varphi_0] = \sum_{k=1}^{\infty}[\varphi_k]$ by Lebesgue's dominated convergence theorem.
Note also that for all $m, n \in \mathbb{N}$ with $1 \leq m < n$ and for every $\omega \in [j\cdot 2^{-m}, j\cdot 2^{-m}+2^{-2n}]$ we have $\sum_{k=m}^{n-1}\varphi_k(\omega) = n - m$. This implies that for any $M > 0$ and any interval $I \subseteq \mathbb{R}$ with nonempty interior there exists an open interval $J \subseteq I$ such that

$$\varphi_0(\omega) > M, \quad \omega \in J. \tag{2.3}$$

To show that φ_0 has the requested property let $\varphi : [0,\infty)\times\mathbb{R} \to \mathbb{R}$ be a Borel measurable function such that $[\varphi(t,\cdot)] = S(t)[\varphi_0]$ for every $t > 0$. Then there exist a countable dense subset $K \subseteq (0,\infty)$ and a Borel nullset $N \subseteq \mathbb{R}$ such that

$$\varphi(t,\omega) = \varphi_0(t+\omega), \quad t \in K,\, \omega \in \mathbb{R}\backslash N. \tag{2.4}$$

Seeking a contradiction we assume that there exists $\omega \in \mathbb{R}\backslash N$ such that $L := \lim_{t\downarrow 0}\varphi(t,\omega)$ exists. Without loss of generality we assume that $\omega \in [0,1)$. If we can construct a nonincreasing sequence $\{t_n\}_{n=1}^{\infty}$ in K such that $\lim_{n\to\infty} t_n = 0$ and $\varphi_0(t_n+\omega) > L+1$ for $n = 1, 2, \cdots$, then we obtain the following contradiction using (2.4):

$$L = \lim_{n\to\infty}\varphi(t_n,\omega) = \lim_{n\to\infty}\varphi_0(t_n+\omega) \geq L+1.$$

This would finish the proof. To construct $\{t_n\}_{n=1}^{\infty}$ we use (2.3) with $M := L+1$. If $I_1 := [\omega, 1)$, then there exist $J_1 \subseteq (\omega, 1)$ and, by density of K, $t_1 \in K$ such that $t_1 + \omega \in J_1$. Now (2.3) implies that $\varphi_0(t_1+\omega) > L+1$. If $I_2 := [\omega, t_1+\omega)$, then there exist $J_2 \subseteq (\omega, t_1+\omega)$ and $t_2 \in K$ such that $t_2+\omega \in J_2$ and hence, $\varphi_0(t_2+\omega) > L+1$. Proceeding like this gives the sequence $\{t_n\}_{n=1}^{\infty}$.

□

Remark 2.2.2 Example 2.2.1 shows more: If φ_0 and φ are as stated in Example 2.2.1, then for every $t_0 > 0$ the mapping $t \mapsto \varphi(t, \omega)$ is discontinuous at t_0 for almost every $\omega \in \mathbb{R}$.

2.3 Versions of analytic functions in $L^p(\Omega; X)$

The main result of this section is that every analytic function with values in $L^p(\Omega; X)$ has an analytic version with values in X. We start with a lemma that is essentially a result of E.M. Stein, rewritten for our convenience, see [Ste70, Lemma, page 72].

We use the notation

$$B(z_0; r) := \big\{ z \in \mathbb{C};\ |z - z_0| < r \big\}, \quad z_0 \in \mathbb{C},\ r > 0.$$

Lemma 2.3.1 *Let $(X, \|\cdot\|)$ be a complex Banach space and $(\Omega, \mathcal{F}, \mu)$ a finite measure space. Let $z_0 \in \mathbb{C}$, $r > 0$, and $\Sigma := B(z_0; r)$. Let $\Phi : \Sigma \to L^1(\Omega; X)$ be a function with an analytic extension to a neighborhood of $\overline{\Sigma}$. Then there exists a function $\varphi : \Sigma \times \Omega \to X$ with the following properties:*

(i) *φ is strongly measurable;*

(ii) *The mapping $z \mapsto \varphi(z, \omega)$ is analytic in Σ for every $\omega \in \Omega$;*

(iii) *For $j = 0, 1, \cdots$ it holds that*

$$\left[\frac{\partial^j}{\partial z^j} \varphi(z, \cdot) \right] = \frac{\mathrm{d}^j}{\mathrm{d}z^j} \Phi(z), \quad z \in \Sigma.$$

PROOF: Without loss of generality we assume that $z_0 = 0$. Firstly, we construct $\varphi : \Sigma \times \Omega \to X$. Since Φ is analytic in a neighborhood of $\overline{\Sigma}$, there exists a sequence $\{C_k\}_{k=0}^{\infty}$ in $L^1(\Omega; X)$ such that

$$\Phi(z) = \sum_{k=0}^{\infty} \frac{z^k}{k!} C_k, \quad z \in \overline{\Sigma}. \tag{2.5}$$

Moreover, the power series in (2.5) has radius of convergence larger than r so that

$$\sum_{k=0}^{\infty} \frac{r^k}{k!} \|C_k\|_{L^1(\Omega;X)} < \infty. \tag{2.6}$$

For $k = 0, 1, \cdots$ we choose a representative $c_k : \Omega \to X$ of the equivalence class C_k. Using Fubini's theorem we obtain

$$\int_\Omega \sum_{k=0}^{\infty} \frac{r^k}{k!} \|c_k(\omega)\| \, \mu(\mathrm{d}\omega) = \sum_{k=0}^{\infty} \frac{r^k}{k!} \int_\Omega \|c_k(\omega)\| \, \mu(\mathrm{d}\omega) = \sum_{k=0}^{\infty} \frac{r^k}{k!} \|C_k\|_{L^1(\Omega;X)}.$$

Combined with (2.6) this implies that there exists a nullset $N \subseteq \Omega$ such that

$$\sum_{k=0}^{\infty} \frac{r^k}{k!} \|c_k(\omega)\| < \infty, \quad \omega \in \Omega \backslash N. \tag{2.7}$$

Now we define $\varphi : \Sigma \times \Omega \to X$ by

$$\varphi(z, \omega) := \begin{cases} \sum_{k=0}^{\infty} \frac{z^k}{k!} c_k(\omega), & z \in \Sigma,\ \omega \in \Omega \backslash N, \\ 0, & z \in \Sigma,\ \omega \in N. \end{cases}$$

Note that φ is well-defined by (2.7) and that the mapping $z \mapsto \varphi(z, \omega)$ is analytic in Σ for every $\omega \in \Omega$. Thus, φ has property (ii).
Secondly, we show that φ has property (i). For $k = 0, 1, \cdots$ the mapping $z \mapsto \frac{z^k}{k!}$ is Borel measurable on Σ. Furthermore, the mapping $\omega \mapsto c_k(\omega)$ is strongly measurable on Ω, since C_k belongs to $L^1(\Omega; X)$. Therefore the mapping $(z, \omega) \mapsto \frac{z^k}{k!} c_k(\omega)$ is strongly measurable on $\Sigma \times \Omega$ for $k = 0, 1, \cdots$. Hence, the mapping $(z, \omega) \mapsto \sum_{k=0}^{\infty} \frac{z^k}{k!} c_k(\omega)$ is strongly measurable on $\Sigma \times \Omega \backslash N$ as pointwise limit of finite sums. It follows that φ is strongly measurable on $\Sigma \times \Omega$.
Thirdly, to show that φ has property (iii) we fix an arbitrary $j \in \mathbb{N} \cup \{0\}$ and observe that

$$\frac{\partial^j}{\partial z^j} \varphi(z, \omega) = \sum_{k=0}^{\infty} \frac{z^k}{k!} c_{k+j}(\omega), \quad z \in \Sigma,\ \omega \in \Omega \backslash N.$$

Let $\varphi_{j,n} : \Sigma \times \Omega \to X$ for $n = 1, 2, \cdots$ be defined by

$$\varphi_{j,n}(z, \omega) := \begin{cases} \sum_{k=0}^{n} \frac{z^k}{k!} c_{k+j}(\omega), & z \in \Sigma,\ \omega \in \Omega \backslash N, \\ 0, & z \in \Sigma,\ \omega \in N. \end{cases}$$

Now we also fix an arbitrary $z \in \Sigma$. On the one hand we have using (2.5),

$$\lim_{n \to \infty} [\varphi_{j,n}(z, \cdot)] = \lim_{n \to \infty} \sum_{k=0}^{n} \frac{z^k}{k!} C_{k+j} = \sum_{k=0}^{\infty} \frac{z^k}{k!} C_{k+j} = \frac{\mathrm{d}^j}{\mathrm{d}z^j} \Phi(z), \tag{2.8}$$

where the convergence is in $L^1(\Omega; X)$. Note that the power series in (2.8) has radius of convergence larger than r. On the other hand we have

$$\lim_{n \to \infty} \varphi_{j,n}(z, \omega) = \frac{\partial^j}{\partial z^j} \varphi(z, \omega), \quad \omega \in \Omega,$$

where the convergence is in X. Since we also have that for $n = 1, 2, \cdots$,

$$\|\varphi_{j,n}(z, \omega)\| \leq \sum_{k=0}^{n} \frac{|z|^k}{k!} \|c_{k+j}(\omega)\| \leq \sum_{k=0}^{\infty} \frac{r^k}{k!} \|c_{k+j}(\omega)\|, \quad \omega \in \Omega \backslash N,$$

and, by Fubini's theorem and (2.5),

$$\int_{\Omega\setminus N}\sum_{k=0}^{\infty}\frac{r^k}{k!}\|c_{k+j}(\omega)\|\,\mu(\mathrm{d}\omega)=\sum_{k=0}^{\infty}\frac{r^k}{k!}\|C_{k+j}\|_{L^1(\Omega;X)}<\infty,$$

we can apply Lebesgue's dominated convergence theorem to obtain

$$\lim_{n\to\infty}\int_{\Omega}\left\|\varphi_{j,n}(z,\omega)-\frac{\partial^j}{\partial z^j}\varphi(z,\omega)\right\|\,\mu(\mathrm{d}\omega)=0.$$

It follows that

$$\lim_{n\to\infty}\left[\varphi_{j,n}(z,\cdot)\right]=\left[\frac{\partial^j}{\partial z^j}\varphi(z,\cdot)\right], \tag{2.9}$$

where the convergence is in $L^1(\Omega;X)$. Now (2.8) and (2.9) show that φ has property (iii). □

We use Lemma 2.3.1 to prove that the same result holds for any open set $\Sigma\subseteq\mathbb{C}$.

Lemma 2.3.2 *Let X be a complex Banach space, $(\Omega,\mathcal{F},\mu)$ a finite measure space, and $\Sigma\subseteq\mathbb{C}$ an open subset. Let $\Phi:\Sigma\to L^1(\Omega;X)$ be an analytic function. Then there exists a function $\varphi:\Sigma\times\Omega\to X$ with the properties (i), (ii), and (iii) stated in Lemma 2.3.1.*

PROOF: As Σ is open in $\mathbb{C}$ it can be covered by countably many open balls. Let $\{z_k\}_{k=1}^{\infty}$ and $\{r_k\}_{k=1}^{\infty}$ be a sequence in Σ respectively in $(0,\infty)$ such that $\overline{B(z_k;r_k)}\subseteq\Sigma$ for $k=1,2,\cdots$ and $\cup_{k=1}^{\infty}B(z_k;r_k)=\Sigma$. It is a result of Lemma 2.3.1 applied to $\Phi|_{B(z_k;r_k)}$ for $k=1,2,\cdots$ that there exists a strongly measurable function $\varphi_k:B(z_k;r_k)\times\Omega\to X$ such that the mapping $z\mapsto\varphi_k(z,\omega)$ is analytic in $B(z_k;r_k)$ for every $\omega\in\Omega$ and, for $j=0,1,\cdots$,

$$\left[\frac{\partial^j}{\partial z^j}\varphi_k(z,\cdot)\right]=\frac{\mathrm{d}^j}{\mathrm{d}z^j}\Phi(z),\quad z\in B(z_k;r_k). \tag{2.10}$$

To construct $\varphi:\Sigma\times\Omega\to X$ we define

$$\begin{aligned}I_z&:=\{k\in\mathbb{N};\ z\in B(z_k;r_k)\}, && z\in\Sigma,\\ N_{k,l,z}&:=\{\omega\in\Omega;\ \varphi_k(z,\omega)\neq\varphi_l(z,\omega)\}, && k,l\in I_z,\ z\in\Sigma.\end{aligned}$$

For every $z\in\Sigma$ and all $k,l\in I_z$ we have that $N_{k,l,z}$ is a nullset in Ω, since $[\varphi_k(z,\cdot)]=\Phi(z)=[\varphi_l(z,\cdot)]$ by (2.10). Now we fix an arbitrary countable dense subset $\Sigma_0\subseteq\Sigma$ and define the nullsets

$$\begin{aligned}N_z&:=\cup_{k,l\in I_z}N_{k,l,z},\quad z\in\Sigma,\\ N&:=\cup_{z\in\Sigma_0}N_z.\end{aligned}$$

Note that if $\omega \in \Omega \backslash N$, then

$$\varphi_k(z, \omega) = \varphi_l(z, \omega), \quad k, l \in I_z, \; z \in \Sigma_0. \tag{2.11}$$

Even more, (2.11) holds for every $z \in \Sigma$ whenever $\omega \in \Omega \backslash N$, since Σ_0 is dense in Σ and the mapping $z \mapsto \varphi_k(z, \omega)$ is continuous in $B(z_k; r_k)$ for $k = 1, 2, \cdots$ and every $\omega \in \Omega$. Finally we define $\varphi : \Sigma \times \Omega \to X$ by

$$\varphi(z, \omega) := \begin{cases} \varphi_k(z, \omega), & z \in \Sigma, \; \omega \in \Omega \backslash N, \; k \in I_z, \\ 0, & z \in \Sigma, \; \omega \in N. \end{cases}$$

By construction, φ is independent of the choice of k. Moreover, it follows from (2.10) that for $j = 0, 1, \cdots$,

$$\left[\frac{\partial^j}{\partial z^j} \varphi(z, \cdot) \right] = \frac{\mathrm{d}^j}{\mathrm{d}z^j} \Phi(z), \quad z \in \Sigma.$$

Thus, φ has property (iii). Now we show that the mapping $z \mapsto \varphi(z, \omega)$ is analytic in Σ for every $\omega \in \Omega$. As this is obvious whenever $\omega \in N$ we fix arbitrary $\omega \in \Omega \backslash N$ and $z_0 \in \Sigma$. For $k \in I_{z_0}$ we observe that z_0 belongs to $B(z_k; r_k)$ and that the mapping $z \mapsto \varphi_k(z, \omega) = \varphi(z, \omega)$ is analytic in $B(z_k; r_k)$. This implies that the mapping $z \mapsto \varphi(z, \omega)$ is analytic in z_0 and hence, φ has property (ii).
Finally, we show that φ is strongly measurable. Let $\{B_k\}_{k=1}^{\infty}$ be the disjoint partition of Σ given by

$$B_k := \begin{cases} B(z_1; r_1), & k = 1, \\ B(z_k; r_k) \setminus \cup_{l=1}^{k-1} B(z_l; r_l), & k = 2, 3, \cdots. \end{cases}$$

Note that $\cup_{k=1}^{\infty} B_k = \Sigma$. Let $\tilde{\varphi}_k : \Sigma \times \Omega \to X$ for $k = 1, 2, \cdots$ be defined by

$$\tilde{\varphi}_k(z, \omega) := \begin{cases} \varphi_k(z, \omega), & z \in B_k, \; \omega \in \Omega \backslash N, \\ 0, & z \in B_k, \; \omega \in N, \\ 0, & z \in \Sigma \backslash B_k, \; \omega \in \Omega. \end{cases}$$

Then $\tilde{\varphi}_k$ is strongly measurable on $\Sigma \times \Omega$, since φ_k is strongly measurable on $B(z_k; r_k) \times \Omega$ for $k = 1, 2, \cdots$. Moreover,

$$\varphi(z, \omega) = \sum_{k=1}^{\infty} \tilde{\varphi}_k(z, \omega), \quad z \in \Sigma, \; \omega \in \Omega.$$

It follows that φ is strongly measurable on $\Sigma \times \Omega$ as pointwise limit of finite sums, that is, φ has property (i). $\square$

The next lemma extends the result of Lemma 2.3.2 to σ-finite measure spaces. The assumptions in the lemma look rather technical, but they essentially concern locally integrable functions. We use the notation

$$M(\Omega; X) := \big\{[\varphi];\ \varphi : \Omega \to X \text{ strongly measurable}\big\}.$$

Lemma 2.3.3 *Let X be a complex Banach space, $(\Omega, \mathcal{F}, \mu)$ a σ-finite measure space, and $\{\Omega_k\}_{k=1}^{\infty}$ a sequence in $\mathcal{F}$ with $\cup_{k=1}^{\infty}\Omega_k = \Omega$. Let $\Sigma \subseteq \mathbb{C}$ be an open subset. Let $\Phi : \Sigma \to M(\Omega; X)$ be such that for $k = 1, 2, \cdots$ the function $\Phi_k : \Sigma \to M(\Omega_k; X)$, given by $\Phi_k(z) := \Phi(z)|_{\Omega_k}$, has $\mathrm{Ran}(\Phi_k) \subseteq L^1(\Omega_k; X)$, and $\Phi_k : \Sigma \to L^1(\Omega_k; X)$ is analytic.*
Then there exists a function $\varphi : \Sigma \times \Omega \to X$ with the following properties:

(i) *φ is strongly measurable;*

(ii) *$[\varphi(z, \cdot)] = \Phi(z)$ for every $z \in \Sigma$;*

(iii) *The mapping $z \mapsto \varphi(z, \omega)$ is analytic in Σ for every $\omega \in \Omega$;*

(iv) *For $j = 0, 1, \cdots$ and $k = 1, 2, \cdots$ it holds that*

$$\left[\frac{\partial^j}{\partial z^j}\varphi(z, \cdot)|_{\Omega_k}\right] = \frac{\mathrm{d}^j}{\mathrm{d}z^j}\Phi_k(z), \quad z \in \Sigma.$$

PROOF: Without loss of generality we assume that $\{\Omega_k\}_{k=1}^{\infty}$ is pairwise disjoint with $\mu(\Omega_k) < \infty$ for $k = 1, 2, \cdots$. It is a result of Lemma 2.3.2 applied to Φ_k for $k = 1, 2, \cdots$ that there exists a strongly measurable function $\varphi_k : \Sigma \times \Omega_k \to X$ such that the mapping $z \mapsto \varphi_k(z, \omega)$ is analytic in Σ for every $\omega \in \Omega_k$, and, for $j = 0, 1, \cdots$,

$$\left[\frac{\partial^j}{\partial z^j}\varphi_k(z, \cdot)\right] = \frac{\mathrm{d}^j}{\mathrm{d}z^j}\Phi_k(z), \quad z \in \Sigma. \tag{2.12}$$

We define $\varphi : \Sigma \times \Omega \to X$ by

$$\varphi(z, \omega) := \sum_{k=1}^{\infty} \varphi_k(z, \omega)\mathbf{1}_{\Omega_k}(\omega), \quad z \in \Sigma,\ \omega \in \Omega.$$

Then φ is well-defined since $\{\Omega_k\}_{k=1}^{\infty}$ is pairwise disjoint. Note that $\varphi(z, \cdot)|_{\Omega_k} = \varphi_k(z, \cdot)$ for $k = 1, 2, \cdots$ and every $z \in \Sigma$. This implies that φ has properties (i) and (iii). Moreover, it is a result of (2.12) that φ has property (iv). In particular, for $k = 1, 2, \cdots$ we have

$$\big[\varphi(z, \cdot)|_{\Omega_k}\big] = \Phi_k(z, \cdot) = \Phi(z)|_{\Omega_k}, \quad z \in \Sigma.$$

As $\cup_{k=1}^{\infty}\Omega_k = \Omega$ it follows that φ has property (ii). □

We remark that if the range of $\Phi : \Sigma \to M(\Omega; X)$ is contained in $L^p(\Omega; X)$ for some $p \in [1, \infty)$, and if $\Phi : \Sigma \to L^p(\Omega; X)$ is analytic, then Φ satisfies the assumptions in Lemma 2.3.3. This gives the main result of this section.

Theorem 2.3.4 *Let X be a complex Banach space, $(\Omega, \mathcal{F}, \mu)$ a σ-finite measure space, and $\Sigma \subseteq \mathbb{C}$ an open subset. Let $\Phi : \Sigma \to L^p(\Omega; X)$ be an analytic function for some $p \in [1, \infty)$. Then there exists a function $\varphi : \Sigma \times \Omega \to X$ with the following properties:*

(i) *φ is strongly measurable;*

(ii) *The mapping $z \mapsto \varphi(z, \omega)$ is analytic in Σ for every $\omega \in \Omega$;*

(iii) *For $j = 0, 1, \cdots$ it holds that*

$$\left[\frac{\partial^j}{\partial z^j}\varphi(z, \cdot)\right] = \frac{\mathrm{d}^j}{\mathrm{d}z^j}\Phi(z), \quad z \in \Sigma.$$

2.4 Versions of solutions to Cauchy problems in L^p-spaces

In this section we use the main result of Section 2.3 to solve the Cauchy problem (2.1). Let $(X, \|\cdot\|)$ be a complex Banach space, $(\Omega, \mathcal{F}, \mu)$ a σ-finite measure space, and $p \in [1, \infty)$. We assume that $\mathcal{A}$ satisfies the following hypothesis:

Hypothesis 2.4.1 The linear operator $\mathcal{A} : \mathrm{D}(\mathcal{A}) \subseteq \mathcal{L}^p(\Omega; X) \to \mathcal{L}^p(\Omega; X)$ has the following properties:

(i) If φ_1 belongs to $\mathrm{D}(\mathcal{A})$ and if φ_2 is in $\mathcal{L}^p(\Omega; X)$ with $[\varphi_2] = [\varphi_1]$, then φ_2 belongs to $\mathrm{D}(\mathcal{A})$ and $[\mathcal{A}\varphi_2] = [\mathcal{A}\varphi_1]$;

(ii) If $\{\varphi_n\}_{n=1}^{\infty}$ is a sequence in $\mathrm{D}(\mathcal{A})$, and if there exist φ, ψ in $\mathcal{L}^p(\Omega; X)$ and a nullset $N_0 \subseteq \Omega$ such that

$$\begin{aligned}
\lim_{n\to\infty} \varphi_n(\omega) &= \varphi(\omega), \quad \omega \in \Omega \backslash N_0, \\
\lim_{n\to\infty} (\mathcal{A}\varphi_n)(\omega) &= \psi(\omega), \quad \omega \in \Omega \backslash N_0, \\
\lim_{n\to\infty} [\varphi_n] &= [\varphi], \\
\lim_{n\to\infty} [\mathcal{A}\varphi_n] &= [\psi],
\end{aligned}$$

where the convergence in the first two lines is in X and in the last two lines in $L^p(\Omega; X)$, then φ belongs to $\mathrm{D}(\mathcal{A})$ and $(\mathcal{A}\varphi)(\omega) = \psi(\omega)$ for every $\omega \in \Omega \backslash N_0$.

We remark that if $\mathcal{A}$ satisfies Hypothesis 2.4.1, then we can define a linear operator $A : \mathrm{D}(A) \subseteq L^p(\Omega; X) \to L^p(\Omega; X)$ by

$$\mathrm{D}(A) := \{\Phi \in L^p(\Omega; X);\ \text{there exists } \varphi \text{ in } \mathrm{D}(\mathcal{A}) \text{ such that } [\varphi] = \Phi\},$$

$$A\Phi := [\mathcal{A}\varphi], \quad \Phi \in \mathrm{D}(A). \tag{2.13}$$

This operator is well-defined by Hypothesis 2.4.1(i). Even more, A is a closed operator.

Theorem 2.4.2 *Let $\mathcal{A}$ satisfy Hypothesis 2.4.1. Let A, defined by (2.13), be the infinitesimal generator of a strongly continuous semigroup $\{S(t)\}_{t\geq 0}$ on $L^p(\Omega; X)$ that is analytic in $\Sigma := \{z \in \mathbb{C}\backslash\{0\};\ |\operatorname{Arg}(z)| < \vartheta\}$ for some $\vartheta \in (0, \pi]$. Let φ_0 belong to $\mathcal{L}^p(\Omega; X)$ and let $\Phi : \Sigma \cup \{0\} \to L^p(\Omega; X)$ be defined by*

$$\Phi(z) := S(z)[\varphi_0], \quad z \in \Sigma.$$

Then there exist a function $\varphi : \Sigma \times \Omega \to X$ and a nullset $N \subseteq \Omega$ with the following properties:

(i) *φ is strongly measurable;*

(ii) *$[\varphi(z, \cdot)] = \Phi(z)$ for every $z \in \Sigma$;*

(iii) *The mapping $z \mapsto \varphi(z, \omega)$ is analytic in Σ for every $\omega \in \Omega$;*

(iv) *$\varphi(z, \cdot)$ belongs to $\mathrm{D}(\mathcal{A})$ for every $z \in \Sigma$, and*

$$\frac{\partial}{\partial z}\varphi(z, \omega) = \big(\mathcal{A}\varphi(z, \cdot)\big)(\omega), \quad z \in \Sigma,\ \omega \in \Omega \backslash N;$$

(v) *If φ_0 belongs to $\mathrm{D}(\mathcal{A})$, then $\lim_{t\downarrow 0}\varphi(t, \omega)$ exists for every $\omega \in \Omega \backslash N$, where the convergence is in X, and*

$$\Big[\lim_{t\downarrow 0}\varphi(t, \cdot)\Big] = [\varphi_0].$$

PROOF: Note that Φ is analytic in Σ. By Theorem 2.3.4 there exists a strongly measurable function $\varphi : \Sigma \times \Omega \to X$ such that the mapping $z \mapsto \varphi(z, \omega)$ is analytic in Σ for every $\omega \in \Omega$, and, for $j = 0, 1, \cdots$,

$$\left[\frac{\partial^j}{\partial z^j}\varphi(z, \cdot)\right] = \frac{\mathrm{d}^j}{\mathrm{d}z^j}\Phi(z), \quad z \in \Sigma. \tag{2.14}$$

Thus, φ has properties (i), (ii), and (iii). Now we prove that φ has property (iv). The first step is to show that $\varphi(z,\cdot)$ belongs to $\mathrm{D}(\mathcal{A})$ for every $z \in \Sigma$ and that

$$\left[\frac{\partial}{\partial z}\varphi(z,\cdot)\right] = \big[\mathcal{A}\varphi(z,\cdot)\big], \quad z \in \Sigma. \tag{2.15}$$

We fix an arbitrary $z \in \Sigma$. Since Φ is defined by an analytic semigroup generated by A, we observe that $\Phi(z)$ belongs to $\mathrm{D}(A)$ and $\frac{\mathrm{d}}{\mathrm{d}z}\Phi(z) = A\Phi(z)$. By definition of $\mathrm{D}(A)$ there exists $\tilde{\varphi}(z,\cdot)$ in $\mathrm{D}(\mathcal{A})$ such that $[\tilde{\varphi}(z,\cdot)] = \Phi(z)$ and $[\mathcal{A}\tilde{\varphi}(z,\cdot)] = A\Phi(z)$. As $[\varphi(z,\cdot)] = \Phi(z)$ by property (ii) we have $[\varphi(z,\cdot)] = [\tilde{\varphi}(z,\cdot)]$. Now it follows from Hypothesis 2.4.1(i) that $\varphi(z,\cdot)$ belongs to $\mathrm{D}(\mathcal{A})$ and $[\mathcal{A}\varphi(z,\cdot)] = [\mathcal{A}\tilde{\varphi}(z,\cdot)]$. Therefore we have, using (2.14),

$$\left[\frac{\partial}{\partial z}\varphi(z,\cdot)\right] = \frac{\mathrm{d}}{\mathrm{d}z}\Phi(z) = A\Phi(z) = [\mathcal{A}\tilde{\varphi}(z,\cdot)] = [\mathcal{A}\varphi(z,\cdot)]\,.$$

The second step is to show that there exists a nullset $N_0 \subseteq \Omega$ such that

$$\frac{\partial}{\partial z}\varphi(z,\omega) = \big(\mathcal{A}\varphi(z,\cdot)\big)(\omega), \quad z \in \Sigma,\ \omega \in \Omega\backslash N_0. \tag{2.16}$$

Let $\Sigma_0 \subseteq \Sigma$ be a countable dense subset. We define the sets

$$N_{z_0} := \left\{\omega \in \Omega;\ \frac{\partial}{\partial z}\varphi(z_0,\omega) \neq \big(\mathcal{A}\varphi(z_0,\cdot)\big)(\omega)\right\}, \quad z_0 \in \Sigma_0,$$
$$N_0 := \cup_{z_0\in\Sigma_0} N_{z_0}.$$

Then (2.15) implies that N_{z_0} is a nullset in Ω for every $z_0 \in \Sigma_0$. Hence, N_0 is a nullset as well and we have

$$\frac{\partial}{\partial z}\varphi(z_0,\omega) = \big(\mathcal{A}\varphi(z_0,\cdot)\big)(\omega), \quad z_0 \in \Sigma_0,\ \omega \in \Omega\backslash N_0. \tag{2.17}$$

Let $z \in \Sigma$ be arbitrary. As Σ_{z_0} is dense in Σ there exists a sequence $\{z_n\}_{n=1}^{\infty}$ in Σ_0 such that $\lim_{n\to\infty} z_n = z$. Even more, by property (iii) we have

$$\lim_{n\to\infty} \varphi(z_n,\omega) = \varphi(z,\omega), \quad \omega \in \Omega,$$

and, using (2.17),

$$\lim_{n\to\infty} \big(\mathcal{A}\varphi(z_n,\cdot)\big)(\omega) = \lim_{n\to\infty}\frac{\partial}{\partial z}\varphi(z_n,\omega) = \frac{\partial}{\partial z}\varphi(z,\omega), \quad \omega \in \Omega\backslash N_0,$$

where the convergence in both lines is in X. Moreover, using (2.14), (2.15), (2.17), and the analyticity of Φ, we have

$$\lim_{n\to\infty} [\varphi(z_n,\cdot)] = \lim_{n\to\infty} \Phi(z_n) = \Phi(z) = [\varphi(z,\cdot)],$$

and

$$\begin{aligned}\lim_{n\to\infty}[\mathcal{A}\varphi(z_n,\cdot)] &= \lim_{n\to\infty}\left[\frac{\partial}{\partial z}\varphi(z_n,\cdot)\right] \\ &= \lim_{n\to\infty}\frac{\mathrm{d}}{\mathrm{d}z}\Phi(z_n) = \frac{\mathrm{d}}{\mathrm{d}z}\Phi(z) = \left[\frac{\partial}{\partial z}\varphi(z,\cdot)\right],\end{aligned}$$

where the convergence in both lines is in $L^p(\Omega;X)$. Now it is a result of Hypothesis 2.4.1(ii), with φ_n, φ, and ψ replaced by respectively $\varphi(z_n,\cdot)$, $\varphi(z,\cdot)$, and $\frac{\partial}{\partial z}\varphi(z,\cdot)$, that (2.16) holds. This shows that φ has property (iv) with N replaced by N_0.
Finally, we prove that φ has property (v). Let $\{\Omega_k\}_{k=1}^{\infty}$ be a pairwise disjoint sequence in Ω with $\cup_{k=1}^{\infty}\Omega_k = \Omega$ and $\mu(\Omega_k) < \infty$ for $k = 1,2,\cdots$. We fix an arbitrary $k \in \mathbb{N}$ and define the analytic function $\Phi_k : \Sigma \to L^p(\Omega_k;X)$ by

$$\Phi_k(z) := \Phi(z)|_{\Omega_k}, \quad z \in \Sigma.$$

The first step is to show that there exists a nullset $N_\infty \subseteq \Omega$ such that $\lim_{t\downarrow 0}\varphi(t,\omega)$ exists for every $\omega \in \Omega\backslash N_\infty$, where the convergence is in X. We have, using Fubini's theorem, (2.14), and the continuous embedding of $L^p(\Omega_k;X)$ into $L^1(\Omega_k;X)$,

$$\int_{\Omega_k}\left(\int_0^T\left\|\frac{\partial}{\partial t}\varphi(t,\omega)\right\|\,\mathrm{d}t\right)\mu(\mathrm{d}\omega) = \int_0^T\left\|\frac{\mathrm{d}}{\mathrm{d}t}\Phi_k(t)\right\|_{L^1(\Omega_k;X)}\mathrm{d}t, \quad T>0.$$

We observe that the integral on the right-hand side is finite. Indeed, the mapping $t\mapsto\Phi_k(t) = (S(t)[\varphi_0])|_{\Omega_k}$ is continuously differentiable on $[0,\infty)$, since the semigroup is strongly continuous and φ_0 belongs to $\mathrm{D}(\mathcal{A})$, see [Paz83, page 102, Theorem 1.3]. Therefore there exists a nullset $N_k \subseteq \Omega_k$ such that for every $T>0$

$$\int_0^T\left\|\frac{\partial}{\partial t}\varphi(t,\omega)\right\|\,\mathrm{d}t < \infty, \quad \omega\in\Omega_k\backslash N_k.$$

This implies that

$$\lim_{t\downarrow 0}\left\|\frac{\partial}{\partial t}\varphi(t,\omega)\right\| < \infty, \quad \omega\in\Omega_k\backslash N_k.$$

Hence, $\lim_{t\downarrow 0}\varphi(t,\omega)$ exists in X for every $\omega\in\Omega_k\backslash N_k$. Since k is arbitrary and $\Omega = \cup_{k=1}^{\infty}\Omega_k$, the first step is completed by defining $N_\infty := \cup_{k=1}^{\infty}N_k$.
The second step is to show that

$$\left[\lim_{t\downarrow 0}\varphi(t,\cdot)\right] = [\varphi_0]. \tag{2.18}$$

By definition of Φ and the strong continuity of the semigroup we have

$$\lim_{t\downarrow 0}\Phi(t) = \lim_{t\downarrow 0}S(t)[\varphi_0] = [\varphi_0], \tag{2.19}$$

where the convergence is in $L^p(\Omega; X)$. Let $\{t_n\}_{n=1}^{\infty}$ be a sequence in $(0,\infty)$ such that $\lim_{n\to\infty} t_n = 0$ and let $k \in \mathbb{N}$ be arbitrary. Then (2.19) implies that

$$\lim_{n\to\infty} \Phi_k(t_n) = \left[\varphi_0|_{\Omega_k}\right],$$

where the convergence is in $L^1(\Omega_k; X)$. By (2.14) we also have

$$\Phi_k(t_n) = \left[\varphi(t_n,\cdot)|_{\Omega_k}\right], \quad n = 1, 2, \cdots .$$

Therefore we have

$$\lim_{n\to\infty} \int_{\Omega_k} \left\|\varphi(t_n,\omega) - \varphi_0(\omega)\right\| \mu(\mathrm{d}\omega) = 0,$$

and hence, there exists a subsequence $\{t_{n_j}\}_{j=1}^{\infty}$ such that for almost every $\omega \in \Omega_k$,

$$\lim_{j\to\infty} \varphi(t_{n_j},\omega) = \varphi_0(\omega), \tag{2.20}$$

where the convergence is in X. Since k is arbitrary and $\Omega = \cup_{k=1}^{\infty}\Omega_k$, we have that (2.20) even holds for almost every $\omega \in \Omega$. Thus, (2.18) holds. This shows that φ has property (v) with N replaced by N_∞ and the theorem is proved with $N := N_0 \cup N_\infty$. □

Chapter 3

Scalar linear Volterra equations

In this chapter we provide an analytic semigroup setting for a scalar linear Volterra integrodifferential equation of convolution type with a completely monotonic kernel. This semigroup approach consists of rewriting the Volterra equation into an abstract Cauchy problem in an appropriate infinite dimensional Hilbert space, such that the operator governing this problem generates an analytic semigroup. We then use the solution to the abstract Cauchy problem, together with interpolation methods, to obtain existence and regularity of solutions to the Volterra equation, as well as a representation formula.

3.1 Introduction

We study the scalar linear Volterra integrodifferential equation

$$\begin{aligned} \frac{\mathrm{d}}{\mathrm{d}t}\int_{-\infty}^{t} a(t-s)u(s)\,\mathrm{d}s &= f(t), \quad t>0, \\ u(t) &= u_0(t), \quad t\le 0, \end{aligned} \tag{3.1}$$

where the kernel $a:(0,\infty)\to\mathbb{R}$ belongs to a class $\mathcal{K}$, introduced in Section 3.2, of completely monotonic, locally integrable functions which are singular at zero, and where the functions $u_0:(-\infty,0]\to\mathbb{R}$ and $f:[0,\infty)\to\mathbb{R}$ are at least locally integrable. We consider the following notion of solution:

Definition 3.1.1 A solution to problem (3.1) is a Borel measurable function $u:\mathbb{R}\to\mathbb{R}$ such that

(i) $\int_{-\infty}^{t} a(t-s)|u(s)|\,\mathrm{d}s<\infty$ for every $t\ge 0$;

(ii) The mapping $t\mapsto\int_{-\infty}^{t} a(t-s)u(s)\,\mathrm{d}s$ is absolutely continuous on $[0,T]$ for every $T>0$;

(iii) u satisfies (3.1) for almost every $t \in \mathbb{R}$.

From this definition we immediately obtain that a solution u to problem (3.1) always belongs to $L^1(0,T)$ for every $T > 0$. Indeed, since a is nonincreasing and $a(t) > 0$ for every $t > 0$, we have

$$a(t) \int_0^t |u(s)| \,\mathrm{d}s \le \int_0^t a(t-s)|u(s)| \,\mathrm{d}s < \infty, \quad t > 0.$$

Concerning uniqueness of solutions we have the following result:

Proposition 3.1.2 *Problem (3.1) admits at most one solution.*

PROOF: Let u be a solution to problem (3.1) with both u_0 and f identically zero. Then Definition 3.1.1(ii) and (iii) imply that $\int_0^t a(t-s)u(s)\,\mathrm{d}s = 0$ for every $t > 0$. It follows from Corollary 1.2.2 that $u(t) = 0$ for almost every $t > 0$. □

To study existence and regularity of solutions to problem (3.1) we use an analytic semigroup approach. Heuristically, this approach is explained as follows. Since a is completely monotonic, Bernstein's theorem states that there exists a unique Borel measure ν on $[0,\infty)$ such that

$$a(t) = \int_{[0,\infty)} \mathrm{e}^{-\kappa t}\,\nu(\mathrm{d}\kappa), \quad t > 0. \tag{3.2}$$

With a view to substitute this expression into (3.1) we define the function $\psi : [0,\infty) \times [0,\infty) \to \mathbb{R}$ by

$$\psi(t,\kappa) := \int_{-\infty}^t \mathrm{e}^{-\kappa(t-s)} u(s)\,\mathrm{d}s, \quad t \ge 0,\ \kappa \ge 0,$$

where the function $u : \mathbb{R} \to \mathbb{R}$ is such that (3.1) holds. Note that ψ satisfies the initial value problem

$$\begin{aligned} \tfrac{\partial}{\partial t}\psi(t,\kappa) &= u(t) - \kappa\psi(t,\kappa), \quad t > 0,\ \kappa \ge 0, \\ \psi(0,\kappa) &= \psi_0(\kappa), \qquad\qquad\qquad\quad\ \ \kappa \ge 0, \end{aligned} \tag{3.3}$$

where the initial value $\psi_0 : [0,\infty) \to \mathbb{R}$ is defined by

$$\psi_0(\kappa) := \int_0^\infty \mathrm{e}^{-\kappa s} u_0(-s)\,\mathrm{d}s, \quad \kappa \ge 0.$$

By substituting (3.2) into (3.1) we obtain

$$\begin{aligned} \frac{\mathrm{d}}{\mathrm{d}t} \int_{-\infty}^t a(t-s)u(s)\,\mathrm{d}s &= \frac{\mathrm{d}}{\mathrm{d}t} \int_{-\infty}^t \left(\int_{[0,\infty)} \mathrm{e}^{-\kappa(t-s)}\,\nu(\mathrm{d}\kappa) \right) u(s)\,\mathrm{d}s \\ = \frac{\mathrm{d}}{\mathrm{d}t} \int_{[0,\infty)} \left(\int_{-\infty}^t \mathrm{e}^{-\kappa(t-s)} u(s)\,\mathrm{d}s \right) \nu(\mathrm{d}\kappa) &= \frac{\mathrm{d}}{\mathrm{d}t} \int_{[0,\infty)} \psi(t,\kappa)\,\nu(\mathrm{d}\kappa) \\ = \int_{[0,\infty)} \tfrac{\partial}{\partial t}\psi(t,\kappa)\,\nu(\mathrm{d}\kappa) &= \int_{[0,\infty)} \big(u(t) - \kappa\psi(t,\kappa)\big)\,\nu(\mathrm{d}\kappa), \qquad t > 0. \end{aligned}$$

Therefore, ψ is subject to the constraint

$$\int_{[0,\infty)} \big(u(t) - \kappa\psi(t,\kappa)\big)\, \nu(\mathrm{d}\kappa) = f(t), \quad t > 0. \tag{3.4}$$

We rewrite the initial value problem (3.3) together with the constraint (3.4) into an abstract Cauchy problem in a suitable Hilbert space. To this end we define a Hilbert space H in Section 3.3, along with a linear operator $A : \mathrm{D}(A) \subseteq H \to H$ that generates an analytic semigroup. These are constructed in such a way that when f is identically zero, the initial value problem (3.3) together with the constraint (3.4) rewrites into the homogeneous abstract Cauchy problem

$$\begin{aligned} \tfrac{\mathrm{d}}{\mathrm{d}t}\varphi(t) &= A\varphi(t), \quad t > 0, \\ \varphi(0) &= \psi_0. \end{aligned} \tag{3.5}$$

In Section 3.4 we discuss problem (3.1) with f identically zero and u_0 such that ψ_0 belongs to $\mathrm{D}(A)$. We solve the homogeneous abstract Cauchy problem (3.5) and obtain a solution to problem (3.1) by means of a linear functional acting on $\varphi(t)$. Using interpolation theory we develop some tools in Section 3.5 which will be used to weaken the assumptions on u_0. Section 3.6 then treats problem (3.1) with f identically zero under these weaker assumptions. In Section 3.7 we consider problem (3.1) with f not identically zero. In this case it turns out that we cannot use the homogeneous abstract Cauchy problem (3.5) as a semigroup setting for problem (3.1). Instead, we think of an inhomogeneous abstract Cauchy problem in a larger space than H. With the aid of the tools of Section 3.5 we solve this problem in an appropriate interpolation space and use its solution to obtain a solution to problem (3.1).

3.2 The kernel

In this section we introduce the class $\mathcal{K}$ mentioned in Section 3.1 of kernels a of the linear Volterra equation (3.1). We show that with every kernel $a \in \mathcal{K}$ we can associate a number $\alpha(a) \in [0,1]$. The notation $\alpha(a)$ is motivated by our model problem where $a = g_{1-\alpha}$ for some $\alpha \in (0,1)$; see (1.2) in Example 1.3.3 for the definition of $g_{1-\alpha}$. In this case the Volterra equation (3.1) with u_0 identically zero equals the fractional derivative of order α, and it turns out that $\alpha(g_{1-\alpha}) = \alpha$. Moreover, we show that if $a \in \mathcal{K}$ behaves like $t^{-\alpha}$ when $t \downarrow 0$, then $\alpha(a) = \alpha$.

Definition 3.2.1 The class $\mathcal{K}$ is defined by

$$\mathcal{K} := \big\{a : (0,\infty) \to \mathbb{R};\ a \text{ completely monotonic}, a \in L^1(0,1), a(0+) = +\infty\big\}.$$

Definition 3.2.2 If $a \in \mathcal{K}$, then $\alpha(a) \in \mathbb{R}$ is defined by

$$\alpha(a) := \sup\left\{\rho \in \mathbb{R};\ \int_1^\infty \lambda^{\rho-2}\frac{1}{(\hat{a}(\lambda))^2}\big(\hat{a}(\lambda) + \lambda\tfrac{\mathrm{d}}{\mathrm{d}\lambda}\hat{a}(\lambda)\big)\,\mathrm{d}\lambda < \infty\right\}.$$

Note that $\alpha(a)$ is well-defined, since the integrand is positive by Lemma 1.1.5(iii) and since $\alpha(a)$ is finite by the next lemma.

Lemma 3.2.3 *If $a \in \mathcal{K}$, then $\alpha(a) \in [0,1]$.*

PROOF: We remark that if the integral in Definition 3.2.2 converges for some $\rho \in \mathbb{R}$, then for every $\rho' < \rho$ this integral with ρ replaced by ρ' also converges. Therefore it is sufficient to prove that

(i) If $\rho < 0$, then $\int_1^\infty \lambda^{\rho-2}\frac{1}{(\hat{a}(\lambda))^2}\big(\hat{a}(\lambda) + \lambda\frac{\mathrm{d}}{\mathrm{d}\lambda}\hat{a}(\lambda)\big)\,\mathrm{d}\lambda < \infty$;

(ii) If $\rho > 1$, then $\int_1^\infty \lambda^{\rho-2}\frac{1}{(\hat{a}(\lambda))^2}\big(\hat{a}(\lambda) + \lambda\frac{\mathrm{d}}{\mathrm{d}\lambda}\hat{a}(\lambda)\big)\,\mathrm{d}\lambda = +\infty$.

For (i) we have by Lemma 1.1.5(ii) and (iv) that $\hat{a}(\lambda) + \lambda\frac{\mathrm{d}}{\mathrm{d}\lambda}\hat{a}(\lambda) \leq \hat{a}(\lambda)$ and $\lambda\hat{a}(\lambda) \geq \hat{a}(1)$ for every $\lambda > 1$. This implies that

$$\begin{aligned}&\int_1^\infty \lambda^{\rho-2}\frac{1}{(\hat{a}(\lambda))^2}\big(\hat{a}(\lambda) + \lambda\tfrac{\mathrm{d}}{\mathrm{d}\lambda}\hat{a}(\lambda)\big)\,\mathrm{d}\lambda \\ &\leq \int_1^\infty \lambda^{\rho-1}\frac{1}{\lambda\hat{a}(\lambda)}\,\mathrm{d}\lambda \leq \frac{1}{\hat{a}(1)}\int_1^\infty \lambda^{\rho-1}\,\mathrm{d}\lambda < \infty, \qquad \rho < 0.\end{aligned}$$

For (ii) we observe that it is sufficient to prove that there exist $M > 0$ and $C > 0$ such that

$$\frac{1}{(\hat{a}(\lambda))^2}\big(\hat{a}(\lambda) + \lambda\tfrac{\mathrm{d}}{\mathrm{d}\lambda}\hat{a}(\lambda)\big) \geq C, \quad \lambda > M,$$

or equivalently, using (3.2) and Lemma 1.1.5(i) and (ii),

$$\frac{\int_{[0,\infty)} \frac{\kappa}{(\kappa+\lambda)^2}\,\nu(\mathrm{d}\kappa)}{\left(\int_{[0,\infty)} \frac{1}{\kappa+\lambda}\,\nu(\mathrm{d}\kappa)\right)^2} \geq C, \quad \lambda > M. \tag{3.6}$$

Let $I_1(\lambda) \geq 0$ and $I_2(\lambda) > 0$ for every $\lambda > 0$ be defined by

$$I_1(\lambda) := \int_{[0,1)} \frac{\lambda}{\kappa+\lambda}\,\nu(\mathrm{d}\kappa), \quad I_2(\lambda) := \int_{[1,\infty)} \frac{\lambda}{\kappa+\lambda}\,\nu(\mathrm{d}\kappa).$$

By Lemma 1.1.4(i) and (iii) we have $\lim_{\lambda\to\infty} I_1(\lambda) = \int_{[0,1)} \nu(\mathrm{d}\kappa) < \infty$ and $\lim_{\lambda\to\infty} I_2(\lambda) = +\infty$. Thus for an arbitrary $\varepsilon > 0$ there exists $M > 0$ such that

$$\frac{I_1(\lambda)}{I_2(\lambda)} < \varepsilon, \quad \lambda > M.$$

Let $c > 0$ be given by

$$c := \int_{[1,\infty)} \frac{1}{\kappa}\,\nu(\mathrm{d}\kappa),$$

where well-definedness follows from Lemma 1.1.6(i). Using the Cauchy-Schwarz inequality we obtain that for every $\lambda > M$,

$$\frac{\int_{[0,\infty)} \frac{\kappa}{(\kappa+\lambda)^2}\,\nu(\mathrm{d}\kappa)}{\left(\int_{[0,\infty)} \frac{1}{\kappa+\lambda}\,\nu(\mathrm{d}\kappa)\right)^2} \geq \frac{1}{\left(1+\frac{I_1(\lambda)}{I_2(\lambda)}\right)^2} \frac{\int_{[1,\infty)} \frac{\kappa}{(\kappa+\lambda)^2}\,\nu(\mathrm{d}\kappa)}{\left(\int_{[1,\infty)} \frac{1}{\kappa+\lambda}\,\nu(\mathrm{d}\kappa)\right)^2} > \frac{1}{(1+\varepsilon)^2}\frac{1}{c}.$$

Therefore (3.6) holds with $C := \frac{1}{c(1+\varepsilon)^2}$. □

Remark 3.2.4 There exists a kernel $a \in \mathcal{K}$ with $\alpha(a) = 1$. Indeed, this holds for $a : (0,\infty) \to \mathbb{R}$ given by

$$a(t) := \int_0^\infty \mathrm{e}^{-\kappa t} \ln(\kappa+2)^{-2}\,\mathrm{d}\kappa, \quad t > 0.$$

Lemma 3.2.5 *If $a \in \mathcal{K}$ is such that $a(t) \sim ct^{-\alpha}$ when $t \downarrow 0$, where $c > 0$ and $\alpha \in (0,1)$, then $\alpha(a) = \alpha$.*

PROOF: We shall show that for $j = 0, 1$,

$$(-1)^j \lambda^j \tfrac{\mathrm{d}^j}{\mathrm{d}\lambda^j} \hat{a}(\lambda) \sim c\Gamma(j+1-\alpha)\lambda^{\alpha-1}, \quad \lambda \to \infty. \tag{3.7}$$

This would prove the lemma, since it implies that for every $\rho \in \mathbb{R}$,

$$\lambda^{\rho-2} \frac{1}{(\hat{a}(\lambda))^2} \left(\hat{a}(\lambda) + \lambda \tfrac{\mathrm{d}}{\mathrm{d}\lambda}\hat{a}(\lambda)\right) \sim \frac{\alpha}{c\Gamma(1-\alpha)} \lambda^{\rho-1-\alpha}, \quad \lambda \to \infty,$$

and $\int_1^\infty \lambda^{\rho-1-\alpha}\,\mathrm{d}\lambda$ converges if and only if $\rho < \alpha$. To show that (3.7) holds, let $\varepsilon > 0$. By assumption there exists $\delta > 0$ such that

$$\left| \frac{a(t)}{ct^{-\alpha}} - 1 \right| < \frac{\varepsilon}{2c\Gamma(1-\alpha)}, \quad 0 < t < \delta.$$

Furthermore, let $c_1 := a(\delta) + c\delta^{-\alpha}$ and let $M > 0$ be such that

$$\lambda^{-\alpha} < \frac{\varepsilon}{2c_1}, \quad \lambda > M.$$

Then we have

$$\begin{aligned}
&\left|(-1)^j \lambda^{j+1-\alpha} \tfrac{\mathrm{d}^j}{\mathrm{d}\lambda^j} \hat{a}(\lambda) - c\Gamma(j+1-\alpha)\right| \\
&= \left|\lambda^{j+1-\alpha} \int_0^\infty e^{-\lambda t} \left(t^j a(t) - ct^{j-\alpha}\right) \mathrm{d}t\right| \\
&\le \lambda^{j+1-\alpha} \int_0^\delta e^{-\lambda t} \left|\frac{a(t)}{ct^{-\alpha}} - 1\right| ct^{j-\alpha}\, \mathrm{d}t + \lambda^{j+1-\alpha} \int_\delta^\infty e^{-\lambda t} t^j |a(t) - ct^{-\alpha}|\, \mathrm{d}t \\
&< \frac{\varepsilon}{2\Gamma(1-\alpha)} \lambda^{j+1-\alpha} \int_0^\infty e^{-\lambda t} t^{j-\alpha}\, \mathrm{d}t + c_1 \lambda^{j+1-\alpha} \int_0^\infty e^{-\lambda t} t^j\, \mathrm{d}t \\
&= \frac{\varepsilon \Gamma(j+1-\alpha)}{2\Gamma(1-\alpha)} + c_1 \Gamma(j+1) \lambda^{-\alpha} < \frac{\varepsilon}{2} + \frac{\varepsilon}{2} = \varepsilon, \qquad \lambda > M,\ j = 0, 1,
\end{aligned}$$

and this implies that (3.7) holds. □

3.3 The analytic semigroup and its generator

This section specifies the analytic semigroup setting. We define the Hilbert space H and the linear operator A, and show firstly, that A is the infinitesimal generator of a strongly continuous semigroup and secondly, that this semigroup is analytic. Let a belong to $\mathcal{K}$ and let ν denote the unique Borel measure on $[0,\infty)$ such that

$$a(t) = \int_{[0,\infty)} \mathrm{e}^{-\kappa t}\, \nu(\mathrm{d}\kappa), \quad t > 0.$$

To define the Hilbert space H we fix an arbitrary $\beta > 0$.

Definition 3.3.1 The real Hilbert space $(H, \langle\cdot,\cdot\rangle_H, \|\cdot\|_H)$ of equivalence classes is defined by

$$H := \Big\{\varphi : [0,\infty) \to \mathbb{R}; \varphi \text{ Borel measurable and } \int_{[0,\infty)} |\varphi(\kappa)|^2 (\kappa+\beta)\, \nu(\mathrm{d}\kappa) < \infty\Big\},$$

endowed with the inner product

$$\langle \varphi, \psi \rangle_H := \int_{[0,\infty)} \varphi(\kappa)\psi(\kappa)(\kappa+\beta)\, \nu(\mathrm{d}\kappa), \quad \varphi, \psi \in H.$$

Note that H is independent of β and that for different choices of β the corresponding norms on H are equivalent.

To define the linear operator A we need a functional J.

Definition 3.3.2 The linear functional $J : \mathrm{D}(J) \subseteq H \to \mathbb{R}$ is defined by

$$\mathrm{D}(J) := \{\varphi \in H; \text{ there exists (a unique) } u \in \mathbb{R} \text{ such that } \kappa \mapsto u - \kappa\varphi(\kappa) \in H\},$$

$$J(\varphi) := u, \quad \varphi \in \mathrm{D}(J).$$

Well-definedness of J follows from the next lemma.

Lemma 3.3.3 *For every $\varphi \in H$ there exists at most one $u \in \mathbb{R}$ such that the mapping $\kappa \mapsto u - \kappa\varphi(\kappa)$ belongs to H.*

PROOF: Let φ belong to H. If $u_1, u_2 \in \mathbb{R}$ are such that the mappings $\kappa \mapsto u_1 - \kappa\varphi(\kappa)$ and $\kappa \mapsto u_2 - \kappa\varphi(\kappa)$ both belong to H, then the mapping $\kappa \mapsto u_1 - u_2$ also belongs to H. This implies that $u_1 = u_2$, since the case $u_1 \neq u_2$ would lead to the following contradiction by Lemma 1.1.4(iv):

$$+\infty = |u_1 - u_2|^2 \int_{[0,\infty)} \kappa\, \nu(\mathrm{d}\kappa) \leq \int_{[0,\infty)} |u_1 - u_2|^2(\kappa + \beta)\, \nu(\mathrm{d}\kappa) < \infty.$$

□

Definition 3.3.4 The linear operator $A : \mathrm{D}(A) \subseteq H \to H$ is defined by

$$\mathrm{D}(A) := \left\{\varphi \in \mathrm{D}(J); \int_{[0,\infty)} \big(J(\varphi) - \kappa\varphi(\kappa)\big)\, \nu(\mathrm{d}\kappa) = 0\right\},$$

$$(A\varphi)(\kappa) := J(\varphi) - \kappa\varphi(\kappa), \quad \varphi \in \mathrm{D}(A),\ \kappa \geq 0.$$

The following lemma shows that A is well-defined.

Lemma 3.3.5 *The Hilbert space H is continuously embedded in the real Banach space $L^1([0,\infty),\nu)$ with*

$$\|\varphi\|_{L^1([0,\infty),\nu)} \leq \sqrt{\hat{a}(\beta)}\|\varphi\|_H, \quad \varphi \in H.$$

PROOF: The lemma follows from the Cauchy-Schwarz inequality and Lemma 1.1.5(i). □

We shall apply the Hille-Yosida theorem to prove that A generates a strongly continuous semigroup. For this we need two propositions concerning dissipativity.

Proposition 3.3.6 *The linear operator $A - \frac{\beta}{4}I_H : \mathrm{D}(A) \subseteq H \to H$ is dissipative in $(H, \langle\cdot,\cdot\rangle_H)$.*

PROOF: Using the definition of D(A) and Lemmas 3.3.5 and 1.1.5(i), we have

$$
\begin{aligned}
\langle A\varphi,\varphi\rangle_H = &\int_{[0,\infty)} \big(J(\varphi)-\kappa\varphi(\kappa)\big)\varphi(\kappa)(\kappa+\beta)\,\nu(\mathrm{d}\kappa) - \\
&J(\varphi)\int_{[0,\infty)} \big(J(\varphi)-\kappa\varphi(\kappa)\big)\,\nu(\mathrm{d}\kappa) \\
= &\int_{[0,\infty)} \left(J(\varphi)\frac{\kappa}{\kappa+\beta}-\kappa\varphi(\kappa)\right)\big(\varphi(\kappa)(\kappa+\beta)-J(\varphi)\big)\,\nu(\mathrm{d}\kappa) + \\
&\int_{[0,\infty)} J(\varphi)\frac{\beta}{\kappa+\beta}\varphi(\kappa)(\kappa+\beta)\,\nu(\mathrm{d}\kappa) - \\
&\int_{[0,\infty)} (J(\varphi))^2\frac{\beta}{\kappa+\beta}\,\nu(\mathrm{d}\kappa) \\
= &-\int_{[0,\infty)} \frac{\kappa}{\kappa+\beta}\big(J(\varphi)-(\kappa+\beta)\varphi(\kappa)\big)^2\,\nu(\mathrm{d}\kappa) + \\
&\beta J(\varphi)\int_{[0,\infty)} \varphi(\kappa)\,\nu(\mathrm{d}\kappa) \;-\; \beta\hat{a}(\beta)(J(\varphi))^2 \\
\le &\;\beta\sqrt{\hat{a}(\beta)}|J(\varphi)|\,\|\varphi\|_H - \beta\hat{a}(\beta)(J(\varphi))^2 \\
= &-\beta\big(\sqrt{\hat{a}(\beta)}\,|J(\varphi)| - \tfrac{1}{2}\|\varphi\|_H\big)^2 + \tfrac{\beta}{4}\|\varphi\|_H^2 \;\le\; \tfrac{\beta}{4}\|\varphi\|_H^2, \quad \varphi\in \mathrm{D}(A).
\end{aligned}
$$

This implies that $\langle (A-\frac{\beta}{4}I_H)\varphi,\varphi\rangle_H \le 0$ for every $\varphi\in\mathrm{D}(A)$. □

Proposition 3.3.7 *The linear operator $A-\frac{\beta}{4}I_H : \mathrm{D}(A)\subseteq H \to H$ is m-dissipative in $(H,\langle\cdot,\cdot\rangle_H)$. Moreover, if $\lambda>0$ and φ belongs to H, then $\lambda I_H - A:\mathrm{D}(A)\to H$ is bijective and*

$$
\big((\lambda I_H - A)^{-1}\varphi\big)(\kappa) = \frac{\varphi(\kappa)+u}{\kappa+\lambda}, \quad \kappa\ge 0,
$$

where $u\in\mathbb{R}$ is defined by

$$
u := \frac{1}{\lambda\hat{a}(\lambda)}\int_{[0,\infty)} \frac{\kappa\varphi(\kappa)}{\kappa+\lambda}\,\nu(\mathrm{d}\kappa). \tag{3.8}
$$

In particular,

$$
u = J\big((\lambda I_H - A)^{-1}\varphi\big).
$$

PROOF: To show that $\lambda I_H - A$ is one-to-one we assume that $\psi\in\mathrm{D}(A)$ satisfies $(\lambda I_H - A)\psi = 0$. Then the definition of A implies that $\psi(\kappa) = \frac{J(\psi)}{\kappa+\lambda}$ for every $\kappa\ge 0$, so that by Lemma 1.1.5(i)

$$
0 = \int_{[0,\infty)} \big(J(\psi)-\kappa\psi(\kappa)\big)\,\nu(\mathrm{d}\kappa) = \int_{[0,\infty)} \frac{\lambda J(\psi)}{\kappa+\lambda}\,\nu(\mathrm{d}\kappa) = \lambda\hat{a}(\lambda)J(\psi).
$$

It follows that $J(\psi) = 0$ and hence, $\psi = 0$.
For well-definedness of u we observe that the mapping $\kappa \mapsto \frac{\kappa\varphi(\kappa)}{\kappa+\lambda}$ belongs to $L^1([0,\infty),\nu)$, since φ belongs to $L^1([0,\infty),\nu)$ by Lemma 3.3.5.
Now we show that $\lambda I_H - A$ is onto. Let $\psi : [0,\infty) \to \mathbb{R}$ be defined by

$$\psi(\kappa) := \frac{\varphi(\kappa) + u}{\kappa + \lambda}, \quad \kappa \geq 0.$$

Then ψ belongs to H as the mappings $\kappa \mapsto \frac{\varphi(\kappa)}{\kappa+\lambda}$ and $\kappa \mapsto \frac{1}{\kappa+\lambda}$ both belong to H; the former since φ belongs to H and the latter by Lemma 1.1.5(ii) and (iii). Therefore it follows from the definition of ψ that the mapping $\kappa \mapsto u - \kappa\psi(\kappa) = \lambda\psi(\kappa) - \varphi(\kappa)$ belongs to H, that is, ψ belongs to $\mathrm{D}(J)$ with $J(\psi) = u$. Even more, ψ belongs to $\mathrm{D}(A)$ since we have, using Lemma 1.1.5(i) and (3.8),

$$\begin{aligned}\int_{[0,\infty)} \big(u - \kappa\psi(\kappa)\big)\,\nu(\mathrm{d}\kappa) &= \int_{[0,\infty)} \frac{\lambda u - \kappa\varphi(\kappa)}{\kappa+\lambda}\,\nu(\mathrm{d}\kappa)\\ &= \lambda\hat{a}(\lambda)u - \int_{[0,\infty)} \frac{\kappa\varphi(\kappa)}{\kappa+\lambda}\,\nu(\mathrm{d}\kappa) = 0.\end{aligned}$$

Finally, the definitions of A and ψ imply that $(\lambda I_H - A)\psi = \varphi$. This shows that $\lambda I_H - A$ is onto and the proof is finished. □

Theorem 3.3.8 *The linear operator $A : \mathrm{D}(A) \subseteq H \to H$ is the infinitesimal generator of a strongly continuous semigroup $\{S(t)\}_{t\geq 0}$ of bounded linear operators on H with*

$$\|S(t)\|_{\mathcal{L}(H)} \leq \mathrm{e}^{\frac{\beta}{4}t}, \quad t \geq 0.$$

PROOF: Since $A - \frac{\beta}{4}I_H$ is m-dissipative in $(H, \langle\cdot,\cdot\rangle_H)$, Proposition 1.4.7 states that A is densely defined in H. Therefore the theorem is a result of Theorem 1.4.5 and Proposition 1.4.9. □

To prove analyticity of $\{S(t)\}_{t\geq 0}$ we need the complexifications of H, J, A, and $\{S(t)\}_{t\geq 0}$. We remark that the complexification of $\{S(t)\}_{t\geq 0}$ is a strongly continuous semigroup on the complexification of H, with infinitesimal generator the complexification of A. Also we remark that in this case Proposition 3.3.7 holds for every $\lambda \in \mathbb{C}\backslash\{0\}$ with $\mathrm{Re}\lambda \geq 0$. We start with two lemmas.

Lemma 3.3.9 *Let $\lambda \in \mathbb{C}$ with $\mathrm{Re}\lambda \geq \frac{\beta}{4}$ and let φ belong to H. If u is given by (3.8), then u satisfies*

$$|u| \leq \frac{2}{|\lambda\hat{a}(\lambda)|}\sqrt{\mathrm{Re}\big(\hat{a}(\tfrac{\beta}{4} + \mathrm{i}\,\mathrm{Im}\lambda)\big)}\,\|\varphi\|_H.$$

PROOF: Using the Cauchy-Schwarz inequality and Lemma 1.1.7, we have

$$\begin{aligned}|u| &\leq \frac{1}{|\lambda\hat{a}(\lambda)|}\int_{[0,\infty)}\frac{\kappa+\beta}{|\kappa+\lambda|}|\varphi(\kappa)|\,\nu(\mathrm{d}\kappa)\\ &\leq \frac{1}{|\lambda\hat{a}(\lambda)|}\left(\int_{[0,\infty)}\frac{\kappa+\beta}{|\kappa+\lambda|^2}\,\nu(\mathrm{d}\kappa)\right)^{\frac{1}{2}}\|\varphi\|_H\\ &\leq \frac{1}{|\lambda\hat{a}(\lambda)|}\left(\int_{[0,\infty)}\frac{4(\kappa+\frac{\beta}{4})}{(\kappa+\frac{\beta}{4})^2+(\mathrm{Im}\lambda)^2}\,\nu(\mathrm{d}\kappa)\right)^{\frac{1}{2}}\|\varphi\|_H\\ &= \frac{2}{|\lambda\hat{a}(\lambda)|}\sqrt{\mathrm{Re}\big(\hat{a}(\tfrac{\beta}{4}+\mathrm{i}\,\mathrm{Im}\lambda)\big)}\,\|\varphi\|_H.\end{aligned}$$

□

Lemma 3.3.10 *The resolvent set $\rho(A)$ contains $\Sigma := \{\lambda \in \mathbb{C};\ \mathrm{Re}\lambda \geq \frac{\beta}{4}\}$. In particular,*

$$\left\|(\lambda I_H - A)^{-1}\right\|_{\mathcal{L}(H)} \leq \left(1 + \frac{4\mathrm{Re}\,\hat{a}(\frac{\beta}{4}+\mathrm{i}\,\mathrm{Im}\lambda)}{|\hat{a}(\lambda)|}\right)\frac{1}{|\lambda|},\quad \lambda\in\Sigma.$$

PROOF: Let $\lambda \in \mathbb{C}$ with $\mathrm{Re}\lambda \geq \frac{\beta}{4}$. Since $\lambda I_H - A$ is bijective by Proposition 3.3.7, it suffices to prove the estimate. Let φ belong to H and let u be given by (3.8). Note that the mapping $\kappa \mapsto \frac{1}{\kappa+\lambda}$ belongs to H by Lemma 1.1.5(ii) and (iii). Using Proposition 3.3.7 and Lemma 3.3.9, we obtain

$$\begin{aligned}\left\|(\lambda I_H - A)^{-1}\varphi\right\|_H &= \left\|\frac{\varphi(\underline{\kappa})+u}{\underline{\kappa}+\lambda}\right\|_H \leq \frac{1}{|\lambda|}\|\varphi\|_H + |u|\left\|\frac{1}{\underline{\kappa}+\lambda}\right\|_H\\ &= \frac{1}{|\lambda|}\|\varphi\|_H + |u|\left(\int_{[0,\infty)}\frac{\kappa+\beta}{|\kappa+\lambda|^2}\,\nu(\mathrm{d}\kappa)\right)^{\frac{1}{2}}\\ &\leq \frac{1}{|\lambda|}\|\varphi\|_H + |u|\left(\int_{[0,\infty)}\frac{4(\kappa+\frac{\beta}{4})}{(\kappa+\frac{\beta}{4})^2+(\mathrm{Im}\lambda)^2}\,\nu(\mathrm{d}\kappa)\right)^{\frac{1}{2}}\\ &\leq \frac{1}{|\lambda|}\|\varphi\|_H + \frac{4}{|\lambda\hat{a}(\lambda)|}\mathrm{Re}\,\hat{a}(\tfrac{\beta}{4}+\mathrm{i}\,\mathrm{Im}\lambda)\,\|\varphi\|_H.\end{aligned}$$

□

Theorem 3.3.11 *The semigroup $\{S(t)\}_{t\geq 0}$ is analytic.*

PROOF: It follows from Lemma 3.3.10 that $\rho(A - \frac{\beta}{4}I_H)$ contains the imaginary axis and that for every $\omega \in \mathbb{R}\backslash\{0\}$,

$$\left\|\left(\mathrm{i}\omega I_H - (A - \tfrac{\beta}{4}I_H)\right)^{-1}\right\|_{\mathcal{L}(H)} \leq \left(1 + \frac{4\mathrm{Re}\,\hat{a}(\frac{\beta}{4}+\mathrm{i}\omega)}{|\hat{a}(\frac{\beta}{4}+\mathrm{i}\omega)|}\right)\frac{1}{|\frac{\beta}{4}+\mathrm{i}\omega|} \leq \frac{5}{|\omega|}.$$

Therefore it is a result of Proposition 1.4.12 that $A - \frac{\beta}{4} I_H$ is the infinitesimal generator of an analytic semigroup. Now Proposition 1.4.13 implies that A is the infinitesimal generator of an analytic semigroup. □

For later use we show that $J|_{\mathrm{D}(A)}$ is continuous with respect to the graph norm of A, which is denoted by $\|\cdot\|_A$.

Lemma 3.3.12 *The restriction* $J|_{\mathrm{D}(A)} : (\mathrm{D}(A), \|\cdot\|_A) \to \mathbb{R}$ *is continuous.*

PROOF: Let ψ belong to $\mathrm{D}(A)$ and let $\varphi \in H$ be given by $\varphi := (\beta I_H - A)\psi$. By Proposition 3.3.7 and Lemma 3.3.5 we have

$$\begin{aligned} |J(\psi)| &\leq \frac{1}{\beta \hat{a}(\beta)} \int_{[0,\infty)} \frac{\kappa}{\kappa + \beta} |\varphi(\kappa)| \, \nu(\mathrm{d}\kappa) \\ &\leq \frac{1}{\beta \hat{a}(\beta)} \|\varphi\|_{L^1([0,\infty),\nu)} \leq \frac{1}{\beta \sqrt{\hat{a}(\beta)}} \big\|(\beta I_H - A)\psi\big\|_H. \end{aligned}$$

Since the norm $\psi \mapsto \|(\beta I_H - A)\psi\|_H$ on $\mathrm{D}(A)$ is equivalent to the graph norm of A, the lemma is proved. □

In Section 3.5 we need the adjoint operator A^* of A. The following results give a characterization of A^* and its resolvent.

Lemma 3.3.13 *The adjoint operator* $A^* : \mathrm{D}(A^*) \subseteq H \to H$ *is given by*

$$\mathrm{D}(A^*) = \left\{ \sigma \in \mathrm{D}(J); \int_{[0,\infty)} \big(J(\sigma) - (\kappa + \beta)\sigma(\kappa)\big) \, \nu(\mathrm{d}\kappa) = 0 \right\},$$

$$(A^*\sigma)(\kappa) = J(\sigma) \frac{\kappa}{\kappa + \beta} - \kappa\sigma(\kappa), \quad \sigma \in \mathrm{D}(A^*), \, \kappa \geq 0.$$

PROOF: Let σ belong to H. By definition of the adjoint operator A^*, σ belongs to $\mathrm{D}(A^*) = \mathrm{D}(\beta I_H - A^*)$ if there exists $\tau \in H$ such that

$$\big\langle (\beta I_H - A)\psi, \sigma \big\rangle_H = \langle \psi, \tau \rangle_H, \quad \psi \in \mathrm{D}(A),$$

or equivalently by Proposition 3.3.7, such that

$$\langle \varphi, \sigma \rangle_H = \big\langle (\beta I_H - A)^{-1} \varphi, \tau \big\rangle_H, \quad \varphi \in H. \tag{3.9}$$

We fix an arbitrary $\varphi \in H$ and let $u \in \mathbb{R}$ be given by

$$u := \frac{1}{\beta \hat{a}(\beta)} \int_{[0,\infty)} \frac{\kappa \varphi(\kappa)}{\kappa + \beta} \nu(\mathrm{d}\kappa).$$

Using Proposition 3.3.7 we have that (3.9) holds if and only if

$$\int_{[0,\infty)} \varphi(\kappa)\sigma(\kappa)(\kappa+\beta)\,\nu(\mathrm{d}\kappa) = \int_{[0,\infty)} (\varphi(\kappa)+u)\tau(\kappa)\,\nu(\mathrm{d}\kappa),$$

if and only if

$$\begin{aligned}&\int_{[0,\infty)} \left(\sigma(\kappa) - \frac{\tau(\kappa)}{\kappa+\beta}\right)\varphi(\kappa)(\kappa+\beta)\,\nu(\mathrm{d}\kappa)\\ &\quad= \frac{1}{\beta\hat{a}(\beta)}\left(\int_{[0,\infty)} \tau(\kappa)\,\nu(\mathrm{d}\kappa)\right)\int_{[0,\infty)} \frac{\kappa}{(\kappa+\beta)^2}\varphi(\kappa)(\kappa+\beta)\,\nu(\mathrm{d}\kappa),\end{aligned}$$

if and only if

$$\left\langle \varphi(\underline{\kappa}), \sigma(\underline{\kappa}) - \frac{\tau(\underline{\kappa})}{\underline{\kappa}+\beta} - \frac{1}{\beta\hat{a}(\beta)}\left(\int_{[0,\infty)} \tau(\kappa)\,\nu(\mathrm{d}\kappa)\right)\frac{\underline{\kappa}}{(\underline{\kappa}+\beta)^2}\right\rangle_H = 0.$$

Since $\varphi \in H$ is arbitrary, this holds if and only if

$$\tau(\kappa) = (\kappa+\beta)\sigma(\kappa) - K\frac{\kappa}{\kappa+\beta}, \quad \kappa \geq 0, \tag{3.10}$$

where, using (3.10) and Lemma 1.1.5(i), $K \in \mathbb{R}$ is such that

$$\begin{aligned}K &= \frac{1}{\beta\hat{a}(\beta)}\int_{[0,\infty)} \tau(\kappa)\,\nu(\mathrm{d}\kappa)\\ &= \frac{1}{\beta\hat{a}(\beta)}\int_{[0,\infty)} \left((\kappa+\beta)\sigma(\kappa) - K + K\frac{\beta}{\kappa+\beta}\right)\nu(\mathrm{d}\kappa)\\ &= \frac{1}{\beta\hat{a}(\beta)}\int_{[0,\infty)} \big((\kappa+\beta)\sigma(\kappa) - K\big)\,\nu(\mathrm{d}\kappa) + K,\end{aligned}$$

or equivalently, where $K \in \mathbb{R}$ satisfies

$$\int_{[0,\infty)} \big(K - (\kappa+\beta)\sigma(\kappa)\big)\,\nu(\mathrm{d}\kappa) = 0. \tag{3.11}$$

Since σ and the mapping $\kappa \mapsto \frac{1}{\kappa+\beta}$ both belong to H, it follows from (3.10) that τ belongs to H if and only if the mapping $\kappa \mapsto \kappa\sigma(\kappa) - K$ belongs to H. We conclude that σ belongs to $\mathrm{D}(A^*)$ if and only if there exists $K \in \mathbb{R}$ such that the mapping $\kappa \mapsto K - \kappa\sigma(\kappa)$ belongs to H and (3.11) holds, that is, σ belongs to $\mathrm{D}(J)$ and satisfies

$$\int_{[0,\infty)} \big(J(\sigma) - (\kappa+\beta)\sigma(\kappa)\big)\,\nu(\mathrm{d}\kappa) = 0.$$

Finally, the definition of A^* and (3.10) imply that

$$(A^*\sigma)(\kappa) = \beta\sigma(\kappa) - \tau(\kappa) = J(\sigma)\frac{\kappa}{\kappa+\beta} - \kappa\sigma(\kappa), \quad \kappa \geq 0.$$

□

Proposition 3.3.14 *If $\lambda > 0$ and τ belongs to H, then $\lambda I_H - A^*$ is bijective and*

$$\big((\lambda I_H - A^*)^{-1}\tau\big)(\kappa) = \frac{1}{\kappa+\lambda}\left(\tau(\kappa) + p\frac{\kappa}{\kappa+\beta}\right), \quad \kappa \geq 0,$$

where $p \in \mathbb{R}$ is defined by

$$p := \frac{1}{\lambda\hat{a}(\lambda)} \int_{[0,\infty)} \frac{\kappa+\beta}{\kappa+\lambda}\tau(\kappa)\,\nu(\mathrm{d}\kappa). \tag{3.12}$$

In particular,

$$p = J\big((\lambda I_H - A^*)^{-1}\tau\big).$$

PROOF: We remark that $\lambda I_H - A^*$ is one-to-one, since $\lambda I_H - A$ is onto by Proposition 3.3.7. Well-definedness of p follows from the observation that the integral in (3.12) is the inner product in H of τ and the mapping $\kappa \mapsto \frac{1}{\kappa+\lambda}$. Note that this mapping belongs to H by Lemma 1.1.5(ii) and (iii).
To show that $\lambda I_H - A^*$ is onto let $\sigma : [0,\infty) \to \mathbb{R}$ be defined by

$$\sigma(\kappa) := \frac{1}{\kappa+\lambda}\left(\tau(\kappa) + p\frac{\kappa}{\kappa+\beta}\right), \quad \kappa \geq 0.$$

Then σ belongs to H as the mappings $\kappa \mapsto \frac{\tau(\kappa)}{\kappa+\lambda}$ and $\kappa \mapsto \frac{\kappa}{(\kappa+\lambda)(\kappa+\beta)}$ both belong to H; the former since τ belongs to H and the latter by Lemma 1.1.5(i). Therefore it follows from the definition of σ and Lemma 1.1.5(i) that the mapping $\kappa \mapsto p - \kappa\sigma(\kappa) = \lambda\sigma(\kappa) - \tau(\kappa) + p\frac{\beta}{\kappa+\beta}$ belongs to H, that is, σ belongs to $\mathrm{D}(J)$ with $J(\sigma) = p$. Even more, σ belongs to $\mathrm{D}(A^*)$ since we have, using Lemma 1.1.5(i) and (3.12),

$$\begin{aligned}\int_{[0,\infty)} \big(p - (\kappa+\beta)\sigma(\kappa)\big)\,\nu(\mathrm{d}\kappa) &= \int_{[0,\infty)} \frac{\lambda p - (\kappa+\beta)\tau(\kappa)}{\kappa+\lambda}\,\nu(\mathrm{d}\kappa) \\ &= \lambda\hat{a}(\lambda)p - \int_{[0,\infty)} \frac{(\kappa+\beta)}{\kappa+\lambda}\tau(\kappa)\,\nu(\mathrm{d}\kappa) = 0.\end{aligned}$$

Now the definitions of A^* and σ imply that $(\lambda I_H - A^*)\sigma = \tau$. This shows that $\lambda I_H - A^*$ is onto and the proof is finished. □

3.4 The homogeneous problem I

We consider the homogeneous Volterra equation

$$\begin{aligned}\frac{\mathrm{d}}{\mathrm{d}t}\int_{-\infty}^{t} a(t-s)u(s)\,\mathrm{d}s = 0, \qquad & t > 0, \\ u(t) = u_0(t), \quad & t \leq 0,\end{aligned} \tag{3.13}$$

where the kernel $a : (0, \infty) \to \mathbb{R}$ belongs to the class $\mathcal{K}$ and where the function $u_0 : (-\infty, 0] \to \mathbb{R}$ satisfies the next hypothesis:

Hypothesis 3.4.1 The function $u_0 : (-\infty, 0] \to \mathbb{R}$ is Borel measurable and has the following properties:

(i) There exist $M_1 > 0$ and $\omega > 0$ such that $|u_0(t)| \leq M_1 \mathrm{e}^{\omega t}$ for every $t \leq 0$;

(ii) There exist $M_2 > 0$ and $\delta > 0$ such that $|u_0(0) - u_0(t)| \leq M_2 |t|$ for every $t \in [-\delta, 0]$;

(iii) $\frac{\mathrm{d}^-}{\mathrm{d}t}\Big|_{t=0} \int_{-\infty}^t a(t-s) u_0(s)\, \mathrm{d}s = 0$.

Note that Hypothesis 3.4.1 does not assume any smoothness of u_0 except at zero. Note also that if u_0 satisfies Hypothesis 3.4.1(i), then u_0 belongs to $L^1(-\infty, 0) \cap L^\infty(-\infty, 0)$. The following lemma gives a different formulation of Hypothesis 3.4.1(iii). We shall give the proof in Section 3.4.3 as it is elementary, but lengthy.

Lemma 3.4.2 *If u_0 satisfies Hypothesis 3.4.1(i) and (ii), then*

$$\begin{aligned} &\frac{\mathrm{d}^-}{\mathrm{d}t}\bigg|_{t=0} \int_{-\infty}^t a(t-s) u_0(s)\, \mathrm{d}s \\ &= u_0(0)\, a(+\infty) + \int_0^\infty \left(-\tfrac{\mathrm{d}}{\mathrm{d}s} a(s)\right)\left(u_0(0) - u_0(-s)\right) \mathrm{d}s. \end{aligned}$$

Following our semigroup approach we consider problem (3.13) in an abstract setting. We study the corresponding homogeneous abstract Cauchy problem in Section 3.4.1, and return to problem (3.13) in Section 3.4.2.

3.4.1 The homogeneous abstract Cauchy problem

In this section we study the homogeneous abstract Cauchy problem

$$\begin{aligned} \tfrac{\mathrm{d}}{\mathrm{d}t}\psi(t) &= A\psi(t), \quad t > 0, \\ \psi(0) &= \psi_0, \end{aligned} \tag{3.14}$$

where $\psi_0 : [0, \infty) \to \mathbb{R}$ is defined by

$$\psi_0(\kappa) := \int_0^\infty \mathrm{e}^{-\kappa s} u_0(-s)\, \mathrm{d}s, \quad \kappa \geq 0. \tag{3.15}$$

We assume that u_0 in the definition of ψ_0 satisfies Hypothesis 3.4.1 and prove that ψ_0 belongs to $\mathrm{D}(A)$. Then we state existence and uniqueness of a strict solution to problem (3.14) in the following sense:

Definition 3.4.3 A *strict solution* to problem (3.14) is a function $\psi : [0,\infty) \to H$ such that ψ belongs to $C\big([0,\infty); \mathrm{D}(A)\big) \cap C^1\big([0,\infty); H\big)$, where $\mathrm{D}(A)$ is endowed with the graph norm of A, and such that ψ satisfies (3.14) for every $t \geq 0$.

Lemma 3.4.4 *If u_0 satisfies Hypothesis 3.4.1(i) and (ii), then ψ_0 belongs to $\mathrm{D}(J)$ and $J(\psi_0) = u_0(0)$. In particular, ψ_0 belongs to $\mathrm{D}(A)$ if and only if u_0 also satisfies Hypothesis 3.4.1(iii).*

PROOF: First we observe that ψ_0 belongs to H. Indeed, by Hypothesis 3.4.1(i) we have $|\psi_0(\kappa)| \leq \frac{M_1}{\kappa+\omega}$ for every $\kappa \geq 0$, and the mapping $\kappa \mapsto \frac{1}{\kappa+\omega}$ belongs to H by Lemma 1.1.5(ii) and (iii). To show that ψ_0 belongs to $\mathrm{D}(J)$ we define $u_1, u_2 : (-\infty, 0] \to \mathbb{R}$ and $\psi_1, \psi_2 : [0,\infty) \to \mathbb{R}$ by respectively $u_1(t) := u_0(0)\mathrm{e}^{\omega t}$ and $u_2(t) := u_0(t) - u_0(0)\mathrm{e}^{\omega t}$ for every $t \leq 0$, and

$$\psi_j(\kappa) := \int_0^\infty \mathrm{e}^{-\kappa s} u_j(-s)\,\mathrm{d}s, \quad \kappa \geq 0,\ j = 1,2.$$

Note that $u_0 = u_1 + u_2$ and $\psi_0 = \psi_1 + \psi_2$, and that u_1 and u_2 satisfy Hypothesis 3.4.1(i) and (ii). Therefore ψ_1 and ψ_2 belong to H. Even more, ψ_1 belongs to $\mathrm{D}(J)$ with $J(\psi_1) = u_0(0)$ since $\psi_1(\kappa) = \frac{u_0(0)}{\kappa+\omega}$ for every $\kappa \geq 0$ and, using Lemma 1.1.5(ii) and (iii),

$$\int_{[0,\infty)} \big(u_0(0) - \kappa\psi_1(\kappa)\big)^2 (\kappa+\beta)\,\nu(\mathrm{d}\kappa) = \big(\omega u_0(0)\big)^2 \int_{[0,\infty)} \frac{\kappa+\beta}{(\kappa+\omega)^2}\,\nu(\mathrm{d}\kappa) < \infty.$$

Now we show that ψ_2 belongs to $\mathrm{D}(J)$ with $J(\psi_2) = 0$. Using Fubini's theorem and Lemma 1.1.4(ii) we observe that

$$\begin{aligned}
&\int_{[0,\infty)} \big|0 - \kappa\psi_2(\kappa)\big|^2 (\kappa+\beta)\,\nu(\mathrm{d}\kappa) = \int_{[0,\infty)} |\psi_2(\kappa)|^2 (\kappa^3 + \beta\kappa^2)\,\nu(\mathrm{d}\kappa) \\
&\leq \int_{[0,\infty)} \left(\int_0^\infty \mathrm{e}^{-\kappa s}|u_2(-s)|\,\mathrm{d}s\right)\left(\int_0^\infty \mathrm{e}^{-\kappa t}|u_2(-t)|\,\mathrm{d}t\right)(\kappa^3 + \beta\kappa^2)\,\nu(\mathrm{d}\kappa) \\
&= \int_0^\infty \int_0^\infty |u_2(-s)||u_2(-t)| \left(-\tfrac{\mathrm{d}^3}{\mathrm{d}t^3} a(s+t) + \beta \tfrac{\mathrm{d}^2}{\mathrm{d}t^2} a(s+t)\right) \mathrm{d}s\,\mathrm{d}t.
\end{aligned}$$

We split the last integral into four parts:

$$\int_0^\infty \int_0^\infty = \int_0^\delta \int_0^\delta + \int_\delta^\infty \int_\delta^\infty + \int_0^\delta \int_\delta^\infty + \int_\delta^\infty \int_0^\delta.$$

It is sufficient to show that each integral on the right-hand side is finite. We shall use Hypothesis 3.4.1(i) and (ii) with M_1' and M_2' for u_2, Fubini's theorem, the assumptions on a, and Lemmas 1.1.5(ii), 1.1.6(ii), and 1.1.2(iv). We start with the first integral:

$$\begin{aligned}
&\int_0^\delta \int_0^\delta |u_2(-s)||u_2(-t)| \left(-\tfrac{\mathrm{d}^3}{\mathrm{d}t^3}a(s+t) + \beta\tfrac{\mathrm{d}^2}{\mathrm{d}t^2}a(s+t)\right) \mathrm{d}s\,\mathrm{d}t \\
&\leq (M_2')^2 \int_{s=0}^\delta s \left(\int_{t=0}^\delta t\left(-\tfrac{\mathrm{d}^3}{\mathrm{d}t^3}a(s+t) + \beta\tfrac{\mathrm{d}^2}{\mathrm{d}t^2}a(s+t)\right) \mathrm{d}t\right) \mathrm{d}s \\
&= (M_2')^2 \int_0^\delta s\Big(\delta\big(-\tfrac{\mathrm{d}^2}{\mathrm{d}t^2}a(s+\delta) + \beta\tfrac{\mathrm{d}}{\mathrm{d}t}a(s+\delta)\big) + \\
&\qquad\qquad \tfrac{\mathrm{d}}{\mathrm{d}t}a(s+\delta) - \beta a(s+\delta) - \tfrac{\mathrm{d}}{\mathrm{d}t}a(s) + \beta a(s)\Big)\,\mathrm{d}s \\
&\leq (M_2')^2 \int_0^\delta s\big(-\tfrac{\mathrm{d}}{\mathrm{d}t}a(s) + \beta a(s)\big)\,\mathrm{d}s \\
&\leq (M_2')^2 \left(-\delta a(\delta) + \int_0^\delta a(s)\,\mathrm{d}s + \beta\delta\int_0^\delta a(s)\,\mathrm{d}s\right) < \infty.
\end{aligned}$$

The second integral is finite, since

$$\begin{aligned}
&\int_\delta^\infty \int_\delta^\infty |u_2(-s)||u_2(-t)| \left(-\tfrac{\mathrm{d}^3}{\mathrm{d}t^3}a(s+t) + \beta\tfrac{\mathrm{d}^2}{\mathrm{d}t^2}a(s+t)\right) \mathrm{d}s\,\mathrm{d}t \\
&\leq (M_1')^2 \int_{s=\delta}^\infty \left(\int_{t=\delta}^\infty \left(-\tfrac{\mathrm{d}^3}{\mathrm{d}t^3}a(s+t) + \beta\tfrac{\mathrm{d}^2}{\mathrm{d}t^2}a(s+t)\right) \mathrm{d}t\right) \mathrm{d}s \\
&= (M_1')^2 \left(-\beta a(+\infty) - \tfrac{\mathrm{d}}{\mathrm{d}t}a(2\delta) + \beta a(2\delta)\right) < \infty.
\end{aligned}$$

Finiteness of the third integral, and analogously of the fourth, follows from

$$\begin{aligned}
&\int_{s=0}^\delta \int_{t=\delta}^\infty |u_2(-s)||u_2(-t)| \left(-\tfrac{\mathrm{d}^3}{\mathrm{d}t^3}a(s+t) + \beta\tfrac{\mathrm{d}^2}{\mathrm{d}t^2}a(s+t)\right) \mathrm{d}s\,\mathrm{d}t \\
&\leq M_1'M_2' \int_{s=0}^\delta s \left(\int_{t=\delta}^\infty \left(-\tfrac{\mathrm{d}^3}{\mathrm{d}t^3}a(s+t) + \beta\tfrac{\mathrm{d}^2}{\mathrm{d}t^2}a(s+t)\right) \mathrm{d}t\right) \mathrm{d}s \\
&= M_1'M_2' \int_0^\delta s\left(\tfrac{\mathrm{d}^2}{\mathrm{d}t^2}a(s+\delta) - \beta\tfrac{\mathrm{d}}{\mathrm{d}t}a(s+\delta)\right) \mathrm{d}s \\
&= M_1'M_2' \left(\delta\big(\tfrac{\mathrm{d}}{\mathrm{d}t}a(2\delta) - \beta a(2\delta)\big) - a(2\delta) + a(\delta) + \beta\int_0^\delta a(s+\delta)\,\mathrm{d}s\right) < \infty.
\end{aligned}$$

At this point we have shown that ψ_1 and ψ_2 belong to $\mathrm{D}(J)$ with $J(\psi_1) = u_0(0)$ and $J(\psi_2) = 0$. Since $\psi_0 = \psi_1 + \psi_2$ it follows that ψ_0 belongs to $\mathrm{D}(J)$ and $J(\psi_0) = u_0(0)$.

Finally, using $\nu(\{0\}) = a(+\infty)$ by Lemma 1.1.4(v), Fubini's theorem justified by the first part of this proof, and Lemma 1.1.4(ii), we observe that

$$\begin{aligned}
&\int_{[0,\infty)} \big(J(\psi_0) - \kappa\psi_0(\kappa)\big)\,\nu(\mathrm{d}\kappa) \\
&= \int_{[0,\infty)} \left(1\cdot u_0(0) - \kappa\int_0^\infty \mathrm{e}^{-\kappa s} u_0(-s)\,\mathrm{d}s\right)\nu(\mathrm{d}\kappa) \\
&= u_0(0)\,a(+\infty) + \int_{(0,\infty)} \left(\int_0^\infty \kappa\mathrm{e}^{-\kappa s}\big(u_0(0) - u_0(-s)\big)\,\mathrm{d}s\right)\nu(\mathrm{d}\kappa) \\
&= u_0(0)\,a(+\infty) + \int_0^\infty \big(-\tfrac{\mathrm{d}}{\mathrm{d}s}a(s)\big)\big(u_0(0) - u_0(-s)\big)\,\mathrm{d}s.
\end{aligned}$$

By Lemma 3.4.2 this shows that ψ_0 belongs to $\mathrm{D}(A)$ if and only if u_0 also satisfies Hypothesis 3.4.1(iii). □

Proposition 3.4.5 *If u_0 satisfies Hypothesis 3.4.1, then problem (3.14) admits a unique strict solution ψ. Moreover, ψ has the representation $\psi(t) = S(t)\psi_0$ for every $t \geq 0$ and ψ is real analytic in $(0,\infty)$ with respect to the graph norm of A.*

PROOF: We refer to [Lun95, page 126, Lemma 4.1.6] and use the analyticity of the semigroup combined with Proposition 1.4.29(ii). □

We remark that the strict solution ψ has values in H and hence, $\psi(t)$ is an equivalence class. In the next section we shall need a differentiable version of $\psi(t)$. To prove existence of such a version we apply Theorem 2.4.2 of Chapter 2. We recall that Chapter 2 distinguishes between a function φ and its equivalence class $[\varphi]$. To be consistent in notation we therefore write $[\psi(t)]$ instead of $\psi(t)$, and denote a version of $[\psi(t)]$ by $\underline{\psi}(t,\cdot)$.
We apply Theorem 2.4.2 with $X := \mathbb{R}$, $\Omega := [0,\infty)$, $\mathcal{F} := \mathcal{B}[0,\infty)$, $\mu(\mathrm{d}\kappa) := (\kappa+\beta)\nu(\mathrm{d}\kappa)$, and $p := 2$. Note that $(\Omega,\mathcal{F},\mu)$ is a σ-finite measure space by Lemma 1.1.4(i). Note also that if $N \subseteq [0,\infty)$ is a μ-null set, then N is a ν-null set as well. To verify the conditions of Theorem 2.4.2 we define the space $\mathcal{H}$ by

$$\mathcal{H} := \big\{\varphi : [0,\infty) \to \mathbb{R};\ [\varphi] \in H\big\},$$

and the linear operator $\mathcal{A} : \mathrm{D}(\mathcal{A}) \subseteq \mathcal{H} \to \mathcal{H}$ by

$$\mathrm{D}(\mathcal{A}) := \big\{\varphi \in \mathcal{H};\ [\varphi] \in \mathrm{D}(A)\big\},$$

$$(\mathcal{A}\varphi)(\kappa) := J([\varphi]) - \kappa\varphi(\kappa), \quad \varphi \in \mathrm{D}(\mathcal{A}),\ \kappa \geq 0.$$

From the definition of A and $\mathcal{A}$ it follows that

$$\mathrm{D}(A) = \big\{[\varphi] \in H;\ \text{there exists } \underline{\varphi} \in \mathrm{D}(\mathcal{A}) \text{ such that } [\underline{\varphi}] = [\varphi]\big\}, \tag{3.16}$$

$$A[\varphi] = [\mathcal{A}\underline{\varphi}], \quad [\varphi] \in \mathrm{D}(A). \tag{3.17}$$

Note that this implies that A could have been defined by (2.13).

Lemma 3.4.6 *Operator $\mathcal{A}$ satisfies Hypothesis 2.4.1.*

PROOF: First we show that $\mathcal{A}$ satisfies Hypothesis 2.4.1(i). Let $\varphi_1 \in \mathrm{D}(\mathcal{A})$ and $\varphi_2 \in \mathcal{H}$ be such that $\varphi_1(\kappa) = \varphi_2(\kappa)$ for μ-almost every $\kappa \geq 0$. Then $[\varphi_1]$ belongs to $\mathrm{D}(A)$ by definition of $\mathcal{A}$, and $[\varphi_1] = [\varphi_2]$. This implies that φ_2 belongs to $\mathrm{D}(\mathcal{A})$ and $\varphi_1(\kappa) = \varphi_2(\kappa)$ for ν-almost every $\kappa \geq 0$. In particular, we have $[\mathcal{A}\varphi_2] = [\mathcal{A}\varphi_1]$ since for μ-almost every $\kappa \geq 0$,

$$(\mathcal{A}\varphi_2)(\kappa) = J([\varphi_2]) - \kappa\varphi_2(\kappa) = J([\varphi_1]) - \kappa\varphi_1(\kappa) = (\mathcal{A}\varphi_1)(\kappa).$$

Now we show that $\mathcal{A}$ satisfies Hypothesis 2.4.1(ii). Let $\{\varphi_n\}_{n=1}^{\infty}$ be a sequence in $\mathrm{D}(\mathcal{A})$, let φ and ψ belong to $\mathcal{H}$, and let $N_0 \subseteq [0,\infty)$ be a μ-null set such that

$$\lim_{n\to\infty} \varphi_n(\kappa) = \varphi(\kappa), \qquad \kappa \in [0,\infty)\backslash N_0, \tag{3.18}$$

$$\lim_{n\to\infty} (\mathcal{A}\varphi_n)(\kappa) = \psi(\kappa), \qquad \kappa \in [0,\infty)\backslash N_0, \tag{3.19}$$

$$\lim_{n\to\infty} [\varphi_n] = [\varphi], \tag{3.20}$$

$$\lim_{n\to\infty} [\mathcal{A}\varphi_n] = [\psi], \tag{3.21}$$

where the convergence in the last two lines is in H. From (3.17) and (3.21) it follows that

$$\lim_{n\to\infty} A[\varphi_n] = \lim_{n\to\infty} [\mathcal{A}\varphi_n] = [\psi]. \tag{3.22}$$

Since A is a closed operator, (3.20) and (3.22) imply that $[\varphi]$ belongs to $\mathrm{D}(A)$ and $A[\varphi] = [\psi]$. This has two consequences. Firstly, by (3.16) there exists $\underline{\varphi} \in \mathrm{D}(\mathcal{A})$ such that $[\underline{\varphi}] = [\varphi]$ and hence, φ belongs to $\mathrm{D}(\mathcal{A})$ by Hypothesis 2.4.1(i). Secondly, using (3.22) we have $\lim_{n\to\infty} A[\varphi_n] = A[\varphi]$. Combined with (3.20) we therefore have $\lim_{n\to\infty}[\varphi_n] = [\varphi]$ where the convergence is in $\mathrm{D}(A)$ with respect to the graph norm $\|\cdot\|_A$ of A. As $J|_{\mathrm{D}(A)} : (\mathrm{D}(A), \|\cdot\|_A) \to \mathbb{R}$ is continuous by Lemma 3.3.12, it follows that $\lim_{n\to\infty} J([\varphi_n]) = J([\varphi])$. Together with (3.18) this implies that for every $\kappa \in [0,\infty)\backslash N_0$,

$$\lim_{n\to\infty} (\mathcal{A}\varphi_n)(\kappa) = \lim_{n\to\infty} \big(J([\varphi_n]) - \kappa\varphi_n(\kappa)\big) = J([\varphi]) - \kappa\varphi(\kappa) = (\mathcal{A}\varphi)(\kappa).$$

In combination with (3.19) this shows that

$$\psi(\kappa) = (\mathcal{A}\varphi)(\kappa), \quad \kappa \in [0,\infty)\backslash N_0.$$

Thus $\mathcal{A}$ satisfies Hypothesis 2.4.1(ii) and the lemma is proved. □

Theorem 3.4.7 *Let* $[\psi(t)]$ *be the unique strict solution to problem (3.14). Then there exist a Borel measurable function* $\underline{\psi} : [0,\infty) \times [0,\infty) \to \mathbb{R}$ *and a* ν*-null set* $N \subseteq [0,\infty)$ *with the following properties:*

(i) $[\underline{\psi}(t,\cdot)] = [\psi(t)]$ *for every* $t \geq 0$*;*

(ii) *The mapping* $t \mapsto \underline{\psi}(t,\kappa)$ *is continuous in* $[0,\infty)$ *and real analytic in* $(0,\infty)$*, both for every* $\kappa \in [0,\infty)\backslash N$*;*

(iii) *It holds that*

$$\tfrac{\partial}{\partial t}\underline{\psi}(t,\kappa) = J([\psi(t)]) - \kappa\underline{\psi}(t,\kappa), \quad t > 0,\ \kappa \in [0,\infty)\backslash N;$$

(iv) $\underline{\psi}(0,\kappa) = \psi_0(\kappa)$ *for every* $\kappa \in [0,\infty)\backslash N$*.*

PROOF: We are in position to apply Theorem 2.4.2, thus there exist a Borel measurable function $\tilde{\varphi} : (0,\infty) \times [0,\infty) \to \mathbb{R}$ and a μ-null set $\tilde{N} \subseteq [0,\infty)$ such that the mapping $t \mapsto \tilde{\varphi}(t,\kappa)$ is real analytic in $(0,\infty)$ for every $\kappa \geq 0$, and $\lim_{t\downarrow 0} \tilde{\varphi}(t,\kappa)$ exists with $[\lim_{t\downarrow 0} \tilde{\varphi}(t,\cdot)] = [\psi_0]$ for every $\kappa \in [0,\infty)\backslash\tilde{N}$. Moreover, $\tilde{\varphi}(t,\cdot)$ belongs to $\mathrm{D}(\mathcal{A})$ for every $t > 0$ and we have

$$\big[\tilde{\varphi}(t,\cdot)\big] = [\psi(t)], \quad t > 0, \tag{3.23}$$

$$\tfrac{\partial}{\partial t}\tilde{\varphi}(t,\kappa) = \big(\mathcal{A}\tilde{\varphi}(t,\cdot)\big)(\kappa), \quad t > 0,\ \kappa \in [0,\infty)\backslash\tilde{N}. \tag{3.24}$$

Now we define $\underline{\psi} : [0,\infty) \times [0,\infty) \to \mathbb{R}$ by

$$\underline{\psi}(t,\kappa) := \begin{cases} \tilde{\varphi}(t,\kappa), & t > 0,\ \kappa \in [0,\infty), \\ \lim_{s\downarrow 0} \tilde{\varphi}(s,\kappa), & t = 0,\ \kappa \in [0,\infty)\backslash\tilde{N}, \\ 0, & t = 0,\ \kappa \in \tilde{N}. \end{cases}$$

Then $\underline{\psi}$ is well-defined and Borel measurable. Furthermore, $\underline{\psi}$ has properties (i) and (ii) with N replaced by $\tilde{N}$. Using (3.23) and (3.24) we have for all $t > 0$ and $\kappa \in [0,\infty)\backslash\tilde{N}$,

$$\begin{aligned} \tfrac{\partial}{\partial t}\underline{\psi}(t,\kappa) &= \tfrac{\partial}{\partial t}\tilde{\varphi}(t,\kappa) = \big(\mathcal{A}\tilde{\varphi}(t,\cdot)\big)(\kappa) \\ &= J\big([\tilde{\varphi}(t,\cdot)]\big) - \kappa\tilde{\varphi}(t,\kappa) = J([\psi(t)]) - \kappa\underline{\psi}(t,\kappa). \end{aligned}$$

Hence, $\underline{\psi}$ has property (iii) with N replaced by $\tilde{N}$. To show that $\underline{\psi}$ has property (iv) we observe that

$$[\underline{\psi}(0,\cdot)] = \left[\lim_{s\downarrow 0} \tilde{\varphi}(s,\cdot)\right] = [\psi_0].$$

Thus there exists a μ-null set $N_0 \subseteq [0,\infty)$ such that

$$\underline{\psi}(0,\kappa) = \psi_0(\kappa), \quad \kappa \in [0,\infty)\backslash N_0.$$

Finally, we define the μ-null set, and thus ν-null set, $N := \tilde{N} \cup N_0$ and the theorem is proved. □

3.4.2 The scalar linear Volterra equation

In this section we return to the homogeneous Volterra equation (3.13), which we are able to solve with our semigroup approach and the results of Section 3.4.1.

Theorem 3.4.8 *Let u_0 satisfy Hypothesis 3.4.1. Then problem (3.13) admits a unique solution u. Moreover, u is continuous in $[0,\infty)$, real analytic in $(0,\infty)$, and has the representation*

$$u(t) = \begin{cases} J\big(S(t)\psi_0\big), & t > 0, \\ u_0(t), & t \le 0, \end{cases} \tag{3.25}$$

where ψ_0 is given by

$$\psi_0(\kappa) := \int_0^\infty \mathrm{e}^{-\kappa s} u_0(-s)\,\mathrm{d}s, \quad \kappa \ge 0.$$

PROOF: Since the mapping $t \mapsto S(t)\psi_0$ is continuous in $[0,\infty)$ and real analytic in $(0,\infty)$, both with respect to the graph norm $\|\cdot\|_A$ of A by Proposition 3.4.5, and since $J : (\mathrm{D}(A), \|\cdot\|_A) \to \mathbb{R}$ is continuous by Lemma 3.3.12, it follows that u given by (3.25) is continuous in $[0,\infty)$ and real analytic in $(0,\infty)$. In particular, u is Borel measurable. Using Hypothesis 3.4.1(i), the continuity of u, and Lemma 1.1.5, we observe that

$$\begin{aligned} &\int_{-\infty}^t a(t-s)|u(s)|\,\mathrm{d}s = \int_{-\infty}^0 a(t-s)|u_0(s)|\,\mathrm{d}s + \int_0^t a(t-s)|u(s)|\,\mathrm{d}s \\ &\le M_1 \int_{-\infty}^0 \mathrm{e}^{\omega s} a(t-s)\,\mathrm{d}s + \max\big\{|u(s)|;\, s\in[0,t]\big\} \int_0^t a(s)\,\mathrm{d}s \\ &\le M_1 \mathrm{e}^{\omega t}\hat{a}(\omega) + \max\big\{|u(s)|;\, s\in[0,t]\big\} \int_0^t a(s)\,\mathrm{d}s < \infty, \qquad t \ge 0. \end{aligned}$$

This implies that u satisfies Definition 3.1.1(i). To show that u satisfies Definition 3.1.1(ii) and (iii) we use the complete monotonicity of a and observe that for every $t \ge 0$,

$$\int_{-\infty}^t a(t-s)u(s)\,\mathrm{d}s = \int_{-\infty}^t \left(\int_{[0,\infty)} \mathrm{e}^{-\kappa(t-s)}\,\nu(\mathrm{d}\kappa) \right) u(s)\,\mathrm{d}s. \tag{3.26}$$

Definition 3.1.1(i) implies that we can apply Fubini's theorem to the right-hand side of (3.26) to obtain for every $t \geq 0$,

$$\int_{-\infty}^{t} \left(\int_{[0,\infty)} \mathrm{e}^{-\kappa(t-s)} \, \nu(\mathrm{d}\kappa) \right) u(s) \, \mathrm{d}s = \int_{[0,\infty)} \left(\int_{-\infty}^{t} \mathrm{e}^{-\kappa(t-s)} u(s) \, \mathrm{d}s \right) \nu(\mathrm{d}\kappa). \tag{3.27}$$

We define $\psi : [0,\infty) \to \mathbb{R}$ by $\psi(t) := S(t)\psi_0$ for every $t \geq 0$. Then ψ is the unique strict solution to problem (3.14) by Proposition 3.4.5. By Theorem 3.4.7 there exist a Borel measurable function $\underline{\psi} : [0,\infty) \times [0,\infty) \to \mathbb{R}$ and a ν-null set $N \subseteq [0,\infty)$ such that the mapping $t \mapsto \underline{\psi}(t,\kappa)$ is continuous in $[0,\infty)$ and real analytic in $(0,\infty)$, both for $\kappa \in [0,\infty)\backslash N$. Moreover, $\underline{\psi}(t,\kappa) = (\psi(t))(\kappa)$ for every $t \geq 0$ and ν-almost every $\kappa \geq 0$, and

$$\begin{aligned} \tfrac{\partial}{\partial t}\underline{\psi}(t,\kappa) &= J(\psi(t)) - \kappa\underline{\psi}(t,\kappa), \quad t > 0,\ \kappa \in [0,\infty)\backslash N, \\ \underline{\psi}(0,\kappa) &= \psi_0(\kappa), \qquad\qquad\qquad\qquad\quad \kappa \in [0,\infty)\backslash N. \end{aligned} \tag{3.28}$$

Hence, using the representations of u and ψ and the definition of ψ_0, we have for all $t \geq 0$ and $\kappa \in [0,\infty)\backslash N$,

$$\underline{\psi}(t,\kappa) = \mathrm{e}^{-\kappa t}\psi_0(\kappa) + \int_0^t \mathrm{e}^{-\kappa(t-s)} J(\psi(s)) \, \mathrm{d}s = \int_{-\infty}^{t} \mathrm{e}^{-\kappa(t-s)} u(s) \, \mathrm{d}s,$$

as well as

$$\underline{\psi}(t,\kappa) = \psi_0(\kappa) + \int_0^t \big(J(\psi(s)) - \kappa\underline{\psi}(s,\kappa) \big) \, \mathrm{d}s.$$

Therefore we have for every $t \geq 0$,

$$\begin{aligned} \int_{[0,\infty)} \left(\int_{-\infty}^{t} \mathrm{e}^{-\kappa(t-s)} u(s) \, \mathrm{d}s \right) \nu(\mathrm{d}\kappa) &= \int_{[0,\infty)} \underline{\psi}(t,\kappa) \, \nu(\mathrm{d}\kappa) \\ = \textstyle\int_{[0,\infty)} \psi_0(\kappa) \, \nu(\mathrm{d}\kappa) &+ \textstyle\int_{[0,\infty)} \left(\int_0^t \big(J(\psi(s)) - \kappa\underline{\psi}(s,\kappa) \big) \, \mathrm{d}s \right) \nu(\mathrm{d}\kappa). \end{aligned} \tag{3.29}$$

To be able to apply Fubini's theorem in (3.29) it is sufficient to show that the mapping $s \mapsto \int_{[0,\infty)} \big(J(\psi(s)) - \kappa\underline{\psi}(s,\kappa) \big) \, \nu(\mathrm{d}\kappa)$ is bounded on $[0,t]$ for every $t \geq 0$. Using Lemma 3.3.5 and the fact that ψ is continuous on $[0,\infty)$ with respect to the graph norm $\|\cdot\|_A$ of A, this follows from

$$\begin{aligned} \int_{[0,\infty)} \big| J(\psi(s)) - \kappa\underline{\psi}(s,\kappa) \big| \, \nu(\mathrm{d}\kappa) &= \int_{[0,\infty)} \big| (A\psi(s))(\kappa) \big| \, \nu(\mathrm{d}\kappa) \\ \leq \sqrt{\hat{a}(\beta)} \|\psi(s)\|_A &\leq \sqrt{\hat{a}(\beta)} \max \big\{ \|\psi(s)\|_A ;\ s \in [0,t] \big\}. \end{aligned}$$

Thus we have for every $t \geq 0$,

$$\begin{aligned} \int_{[0,\infty)} \psi_0(\kappa) \, \nu(\mathrm{d}\kappa) &+ \int_{[0,\infty)} \left(\int_0^t \big(J(\psi(s)) - \kappa\underline{\psi}(s,\kappa) \big) \, \mathrm{d}s \right) \nu(\mathrm{d}\kappa) \\ = \textstyle\int_{[0,\infty)} \psi_0(\kappa) \, \nu(\mathrm{d}\kappa) &+ \textstyle\int_0^t \left(\int_{[0,\infty)} \big(J(\psi(s)) - \kappa(\psi(s))(\kappa) \big) \, \nu(\mathrm{d}\kappa) \right) \mathrm{d}s. \end{aligned} \tag{3.30}$$

Combining (3.26), (3.27), (3.29), and (3.30), and recalling that $\psi(s)$ belongs to $\mathrm{D}(A)$ for every $s \geq 0$, we obtain

$$\int_{-\infty}^{t} a(t-s)u(s)\,\mathrm{d}s = \int_{[0,\infty)} \psi_0(\kappa)\,\nu(\mathrm{d}\kappa), \quad t \geq 0.$$

Since the right-hand side above is independent of t it follows that u satisfies Definition 3.1.1(ii) and (iii). □

3.4.3 Proof of Lemma 3.4.2

In this section we prove Lemma 3.4.2 by applying the following proposition, see [Roy88, page 92, Theorem 17].

Proposition 3.4.9 *Let $\{g_n\}_{n=1}^{\infty}$ be a sequence of nonnegative integrable functions that converges almost everywhere to an integrable function g. Let $\{f_n\}_{n=1}^{\infty}$ be a sequence of measurable functions such that $|f_n| \leq g_n$ for $n = 1, 2, \cdots$, and such that the sequence converges almost everywhere to a function f. If $\lim_{n\to\infty} \int g_n = \int g$, then $\lim_{n\to\infty} f_n = \int f$.*

PROOF OF LEMMA 3.4.2: We start with observing that the integral on the right-hand side exists. Indeed, using Hypothesis 3.4.1(i) and (ii), Lemma 1.1.6(ii), and the assumptions on a, we have

$$\begin{aligned}
&\int_0^{\infty} \left|\big(-\tfrac{\mathrm{d}}{\mathrm{d}s}a(s)\big)\big(u_0(0)-u_0(-s)\big)\right| \mathrm{d}s \\
&= \int_0^{\delta} \big(-\tfrac{\mathrm{d}}{\mathrm{d}s}a(s)\big)\left|u_0(0)-u_0(-s)\right| \mathrm{d}s + \int_{\delta}^{\infty} \big(-\tfrac{\mathrm{d}}{\mathrm{d}s}a(s)\big)\left|u_0(0)-u_0(-s)\right| \mathrm{d}s \\
&\leq M_2 \int_0^{\delta} s\big(-\tfrac{\mathrm{d}}{\mathrm{d}s}a(s)\big)\,\mathrm{d}s + 2M_1 \int_{\delta}^{\infty} -\tfrac{\mathrm{d}}{\mathrm{d}s}a(s)\,\mathrm{d}s \\
&\leq -M_2\delta a(\delta) + M_2 \int_0^{\delta} a(s)\,\mathrm{d}s - 2M_1 a(+\infty) + 2M_1 a(\delta) < \infty.
\end{aligned}$$

To prove the equality in the proposition let $u_1, u_2 : (-\infty, 0] \to \mathbb{R}$ be defined by $u_1(t) := u_0(0)\mathrm{e}^{\omega t}$ respectively $u_2(t) := u_0(t) - u_1(t)$. Then u_1 and u_2 satisfy Hypothesis 3.4.1(i) and (ii). Note that $u_0 = u_1 + u_2$, $u_1(0) = u_0(0)$, and $u_2(0) = 0$. First we show that

$$\begin{aligned}
&\frac{\mathrm{d}^-}{\mathrm{d}t}\bigg|_{t=0} \int_{-\infty}^{t} a(t-s)u_1(s)\,\mathrm{d}s \\
&= u_0(0)\,a(+\infty) + \int_0^{\infty} \big(-\tfrac{\mathrm{d}}{\mathrm{d}s}a(s)\big)\big(u_0(0)-u_1(-s)\big)\,\mathrm{d}s.
\end{aligned} \tag{3.31}$$

For the left-hand side of (3.31) we have, using Lemma 1.1.5,

$$\frac{\mathrm{d}^-}{\mathrm{d}t}\bigg|_{t=0} \int_{-\infty}^{t} a(t-s)u_1(s)\,\mathrm{d}s = u_0(0)\,\frac{\mathrm{d}^-}{\mathrm{d}t}\bigg|_{t=0} \int_{0}^{\infty} a(s)\mathrm{e}^{\omega(t-s)}\,\mathrm{d}s$$
$$= u_0(0)\tfrac{\mathrm{d}^-}{\mathrm{d}t}\big|_{t=0}\mathrm{e}^{\omega t}\hat{a}(\omega) = u_0(0)\omega\hat{a}(\omega).$$

For the right-hand side of (3.31) we use Lemma 1.1.6(ii) to obtain

$$\int_0^\infty \big(-\tfrac{\mathrm{d}}{\mathrm{d}s}a(s)\big)\big(u_0(0)-u_1(-s)\big)\,\mathrm{d}s$$
$$= u_0(0)\int_0^\infty \big(-\tfrac{\mathrm{d}}{\mathrm{d}s}a(s)\big)\big(1-\mathrm{e}^{-\omega s}\big)\,\mathrm{d}s$$
$$= u_0(0)\left(-a(+\infty)+\omega\lim_{s\downarrow 0}\left(sa(s)\frac{1-\mathrm{e}^{-\omega s}}{\omega s}\right)+\omega\int_0^\infty a(s)\mathrm{e}^{-\omega s}\,\mathrm{d}s\right)$$
$$= -u_0(0)\,a(+\infty)+u_0(0)\omega\hat{a}(\omega).$$

This implies that (3.31) holds. Now we show that

$$\frac{\mathrm{d}^-}{\mathrm{d}t}\bigg|_{t=0} \int_{-\infty}^{t} a(t-s)u_2(s)\,\mathrm{d}s = \int_0^\infty \big(\tfrac{\mathrm{d}}{\mathrm{d}s}a(s)\big)u_2(-s)\,\mathrm{d}s. \tag{3.32}$$

By definition of the left derivative we have

$$\frac{\mathrm{d}^-}{\mathrm{d}t}\bigg|_{t=0} \int_{-\infty}^{t} a(t-s)u_2(s)\,\mathrm{d}s$$
$$= \lim_{h\downarrow 0}\tfrac{1}{h}\left(\int_{-\infty}^{0} a(-s)u_2(s)\,\mathrm{d}s - \int_{-\infty}^{-h} a(-h-s)u_2(s)\,\mathrm{d}s\right).$$

For an arbitrary $h\in(0,\frac{\delta}{2})$, where δ is given by Hypothesis 3.4.1(ii), we write

$$\frac{1}{h}\left(\int_{-\infty}^{0} a(-s)u_2(s)\,\mathrm{d}s - \int_{-\infty}^{-h} a(-h-s)u_2(s)\,\mathrm{d}s\right)$$
$$= \frac{1}{h}\left(\int_{0}^{\infty} a(s)u_2(-s)\,\mathrm{d}s - \int_{h}^{\infty} a(s-h)u_2(-s)\,\mathrm{d}s\right)$$
$$= \frac{1}{h}\int_0^{2h} a(s)u_2(-s)\,\mathrm{d}s - \frac{1}{h}\int_h^{2h} a(s-h)u_2(-s)\,\mathrm{d}s +$$
$$\int_{2h}^{\delta}\frac{a(s)-a(s-h)}{h}u_2(-s)\,\mathrm{d}s + \int_{\delta}^{\infty}\frac{a(s)-a(s-h)}{h}u_2(-s)\,\mathrm{d}s.$$

Thus (3.32) holds if we can show the following:

$$\lim_{h\downarrow 0}\frac{1}{h}\int_0^{2h} a(s)u_2(-s)\,\mathrm{d}s = 0 = \lim_{h\downarrow 0}\frac{1}{h}\int_h^{2h} a(s-h)u_2(-s)\,\mathrm{d}s. \tag{3.33}$$

$$\lim_{h\downarrow 0}\int_{\delta}^{\infty}\frac{a(s)-a(s-h)}{h}u_2(-s)\,\mathrm{d}s=\int_{\delta}^{\infty}\big(\tfrac{\mathrm{d}}{\mathrm{d}s}a(s)\big)u_2(-s)\,\mathrm{d}s. \tag{3.34}$$

$$\lim_{h\downarrow 0}\int_{2h}^{\delta}\frac{a(s)-a(s-h)}{h}u_2(-s)\,\mathrm{d}s=\int_{0}^{\delta}\big(\tfrac{\mathrm{d}}{\mathrm{d}s}a(s)\big)u_2(-s)\,\mathrm{d}s. \tag{3.35}$$

To show the first equality in (3.33) we observe that by Hypothesis 3.4.1(i) and (ii) there exist $K_1>0$ and $K_2>0$ such that $|u_2(-t)|\le K_1\mathrm{e}^{-\omega t}$ for every $t\ge 0$, and $|u_2(-t)|\le K_2 t$ for every $t\in[0,\delta]$. Therefore we have

$$\left|\frac{1}{h}\int_0^{2h}a(s)u_2(-s)\,\mathrm{d}s\right|\le\frac{1}{h}\int_0^{2h}a(s)K_2 s\,\mathrm{d}s\le 2K_2\int_0^{2h}a(s)\,\mathrm{d}s.$$

This implies the first equality in (3.33). The second equality in (3.33) follows analogously.

For (3.34) and (3.35) we observe that the mean value theorem and the fact that the mapping $t\mapsto -\frac{\mathrm{d}}{\mathrm{d}t}a(t)$ is nonincreasing imply that

$$\big|a(s)-a(s-h)\big|\le h\big(-\tfrac{\mathrm{d}}{\mathrm{d}s}a(s-h)\big),\quad s>h. \tag{3.36}$$

Thus we have for every $s>\delta$,

$$\left|\frac{a(s)-a(s-h)}{h}u_2(-s)\right|\le\big(-\tfrac{\mathrm{d}}{\mathrm{d}s}a(s-h)\big)K_1\mathrm{e}^{-\omega s}\le K_1\big(-\tfrac{\mathrm{d}}{\mathrm{d}s}a(s-\tfrac{\delta}{2})\big),$$

so that (3.34) follows from Lebesgue's dominated convergence theorem.

It remains to show (3.35) for which we shall apply Proposition 3.4.9. Let the functions $f_h, f:(0,\delta)\to\mathbb{R}$ be defined by

$$\begin{aligned} f_h(s)&:=\frac{a(s)-a(s-h)}{h}u_2(-s)\mathbf{1}_{[2h,\delta]}(s), && s\in(0,\delta),\\ f(s)&:=\big(\tfrac{\mathrm{d}}{\mathrm{d}s}a(s)\big)u_2(-s), && s\in(0,\delta).\end{aligned}$$

Then we have $\lim_{h\downarrow 0}f_h(s)=f(s)$ for every $s\in(0,\delta)$. Moreover, using (3.36) and the fact that $|u_2(-s)|\le K_2 s\le 2K_2(s-h)$ for every $s\in[2h,\delta]$, we have

$$|f_h(s)|\le\big(-\tfrac{\mathrm{d}}{\mathrm{d}s}a(s-h)\big)2K_2(s-h)\mathbf{1}_{[2h,\delta]}(s),\quad s\in(0,\delta).$$

Now we define $g_h, g:(0,\delta)\to\mathbb{R}$ by

$$\begin{aligned} g_h(s)&:=2K_2(s-h)\big(-\tfrac{\mathrm{d}}{\mathrm{d}s}a(s-h)\big)\mathbf{1}_{[2h,\delta]}(s), && s\in(0,\delta),\\ g(s)&:=2K_2 s\big(-\tfrac{\mathrm{d}}{\mathrm{d}s}a(s)\big), && s\in(0,\delta).\end{aligned}$$

Thus we have $|f_h(s)|\le g_h(s)$ and $\lim_{h\downarrow 0}g_h(s)=g(s)$ for every $s\in(0,\delta)$. Moreover, using Lemma 1.1.6(ii) we observe that

$$\begin{aligned}\int_0^{\delta}g_h(s)\,\mathrm{d}s&=2K_2\int_{2h}^{\delta}(s-h)\big(-\tfrac{\mathrm{d}}{\mathrm{d}s}a(s-h)\big)\,\mathrm{d}s\\ &=-2K_2(\delta-h)a(\delta-h)+2K_2ha(h)+2K_2\int_{2h}^{\delta}a(s-h)\,\mathrm{d}s<\infty,\end{aligned}$$

and

$$\int_0^\delta g(s)\,\mathrm{d}s = 2K_2 \int_0^\delta s\big(-\tfrac{\mathrm{d}}{\mathrm{d}s}a(s)\big)\,\mathrm{d}s = -2K_2\delta a(\delta) + 2K_2 \int_0^\delta a(s)\,\mathrm{d}s < \infty.$$

The last fact that should be verified before we can apply Proposition 3.4.9 is whether

$$\lim_{h\downarrow 0} \int_0^\delta \big|g_h(s) - g(s)\big|\,\mathrm{d}s = \lim_{h\downarrow 0} \int_0^{2h} g(s)\,\mathrm{d}s + \lim_{h\downarrow 0} \int_{2h}^\delta \big|g_h(s) - g(s)\big|\,\mathrm{d}s = 0.$$

Using Lemma 1.1.6(ii) we have

$$\begin{aligned} \frac{1}{2K_2} \lim_{h\downarrow 0} \int_0^{2h} g(s)\,\mathrm{d}s &= \lim_{h\downarrow 0} \int_0^{2h} s\big(-\tfrac{\mathrm{d}}{\mathrm{d}s}a(s)\big)\,\mathrm{d}s \\ &= \lim_{h\downarrow 0} \left(-2ha(2h) + \int_0^{2h} a(s)\,\mathrm{d}s\right) = 0. \end{aligned}$$

Furthermore, the mean value theorem combined with the triangle inequality implies that for every $s \in [2h, \delta]$,

$$\big|(s-h)\tfrac{\mathrm{d}}{\mathrm{d}s}a(s-h) - s\tfrac{\mathrm{d}}{\mathrm{d}s}a(s)\big| \le h\big(-\tfrac{\mathrm{d}}{\mathrm{d}s}a(s-h) + s\tfrac{\mathrm{d}^2}{\mathrm{d}s^2}a(s-h)\big).$$

Hence, using Lemma 1.1.6(ii) we have

$$\begin{aligned} &\frac{1}{2K_2} \lim_{h\downarrow 0} \int_{2h}^\delta \big|g_h(s) - g(s)\big|\,\mathrm{d}s = \lim_{h\downarrow 0} \int_{2h}^\delta \big|(s-h)\tfrac{\mathrm{d}}{\mathrm{d}s}a(s-h) - s\tfrac{\mathrm{d}}{\mathrm{d}s}a(s)\big|\,\mathrm{d}s \\ &\le \lim_{h\downarrow 0} h \int_{2h}^\delta \big(-\tfrac{\mathrm{d}}{\mathrm{d}s}a(s-h) + s\tfrac{\mathrm{d}^2}{\mathrm{d}s^2}a(s-h)\big)\,\mathrm{d}s \\ &= \lim_{h\downarrow 0} \big(-2ha(\delta-h) + 2ha(h) + h\delta\tfrac{\mathrm{d}}{\mathrm{d}s}\big|_{s=\delta}a(s-h) - 2h^2\tfrac{\mathrm{d}}{\mathrm{d}s}\big|_{s=2h}a(s-h)\big) \\ &= 0. \end{aligned}$$

Now we are in a position to apply Proposition 3.4.9. We obtain that

$$\lim_{h\downarrow 0} \int_0^\delta f_h(s)\,\mathrm{d}s = \int_0^\delta f(s)\,\mathrm{d}s,$$

that is, (3.35) holds and the lemma is proved. □

3.5 Interpolation spaces

In Section 3.4 we have solved problem (3.1) with $a \in \mathcal{K}$ and f identically zero under some assumptions on u_0. To be able to solve this problem under weaker assumptions in Section 3.6, or even with f not identically zero in Section 3.7, we need some tools. In this section we develop these tools using interpolation spaces.

We recall that with every kernel $a \in \mathcal{K}$ we associate a number $\alpha(a) \in [0,1]$ that was introduced in Definition 3.2.2.

3.5.1 Continuous extension of $J|_{\mathrm{D}(A)}$

The goal of this section is to prove that the restriction $J|_{D(A)}$ has a continuous extension to an interpolation space $(H, \mathrm{D}(A))_{\eta,2}$ for some $\eta \in (0,1)$. For this purpose we need a function $\xi : [0,\infty) \to \mathbb{R}$ that belongs to an interpolation space $(H, \mathrm{D}(A^*))_{1-\eta,2}$.

Definition 3.5.1 The function $\xi : [0,\infty) \to \mathbb{R}$ is defined by

$$\xi(\kappa) := \frac{\kappa}{(\kappa+\beta)^2}, \quad \kappa \geq 0.$$

Note that ξ belongs to H by Lemma 1.1.5(i).

Proposition 3.5.2 *The function ξ belongs to $(H, \mathrm{D}(A^*))_{1-\eta,2}$ for every $\eta \in (0,1)$ such that $\eta > \frac{1-\alpha(a)}{2}$.*

PROOF: Let $\eta \in (0,1)$. We show that ξ belongs to $(H, \mathrm{D}(A^*))_{1-\eta,2}$ if and only if

$$\int_1^\infty t^{-1-2\eta} \frac{1}{(\hat{a}(t))^2} \big(\hat{a}(t) + t \tfrac{\mathrm{d}}{\mathrm{d}t}\hat{a}(t)\big)\,\mathrm{d}t < \infty. \tag{3.37}$$

This would prove the proposition. Indeed, if $\eta > \frac{1-\alpha(a)}{2}$, then $1-2\eta < \alpha(a)$, so that (3.37) holds by definition of $\alpha(a)$ and hence, ξ belongs to $(H, \mathrm{D}(A^*))_{1-\eta,2}$. As $\mathrm{D}(A^*) = \mathrm{D}(A^* - \beta I_H)$ with equivalent norms we have by definition of interpolation spaces that ξ belongs to $(H, D(A^*))_{1-\eta,2}$ if

$$\int_c^\infty t^{2(1-\eta)-1} \big\|(A^* - \beta I_H)\big(tI_H - (A^* - \beta I_H)\big)^{-1}\xi\big\|_H^2 \,\mathrm{d}t < \infty$$

for some $c > 1$ sufficiently large. This holds if and only if

$$\int_{c'}^\infty t^{1-2\eta} \big\|(A^* - \beta I_H)(tI_H - A^*)^{-1}\xi\big\|_H^2 \,\mathrm{d}t < \infty, \tag{3.38}$$

for some $c' > 1$ sufficiently large. By Proposition 3.3.14 we have for all $t > 0$ and $\kappa \geq 0$,

$$\begin{aligned}\big((A^* - \beta I_H)(tI_H - A^*)^{-1}\xi\big)(\kappa) &= -\xi(\kappa) + (t-\beta)\frac{1}{\kappa+t}\left(\xi(\kappa) + p\frac{\kappa}{\kappa+\beta}\right) \\ &= \big(p(t-\beta) - 1\big)\frac{\kappa}{(\kappa+t)(\kappa+\beta)},\end{aligned}$$

where, using Lemma 1.1.5(i), $p \in \mathbb{R}$ satisfies

$$\begin{aligned} p &= \frac{1}{t\hat{a}(t)} \int_{[0,\infty)} \frac{\kappa+\beta}{\kappa+t} \xi(\kappa)\, \nu(\mathrm{d}\kappa) = \frac{1}{t\hat{a}(t)} \int_{[0,\infty)} \frac{\kappa}{(\kappa+t)(\kappa+\beta)}\, \nu(\mathrm{d}\kappa) \\ &= \frac{1}{t\hat{a}(t)} \int_{[0,\infty)} \frac{1}{t-\beta} \left(\frac{t}{\kappa+t} - \frac{\beta}{\kappa+\beta} \right) \nu(\mathrm{d}\kappa) \\ &= \frac{1}{t(t-\beta)\hat{a}(t)} \big(t\hat{a}(t) - \beta\hat{a}(\beta)\big) = \frac{1}{t-\beta}\left(1 - \frac{\beta\hat{a}(\beta)}{t\hat{a}(t)}\right). \end{aligned}$$

Thus we have

$$\big((A^* - \beta I_H)(tI_H - A^*)^{-1}\xi\big)(\kappa) = -\frac{\beta\hat{a}(\beta)}{t\hat{a}(t)} \frac{\kappa}{(\kappa+t)(\kappa+\beta)}, \quad t > 0,\ \kappa \geq 0.$$

Using Lemma 1.1.5(vi) we obtain that for every $t > \beta$,

$$\begin{aligned} &\big\|(A^* - \beta I_H)(tI_H - A^*)^{-1}\xi\big\|_H^2 \\ &= \frac{\big(\beta\hat{a}(\beta)\big)^2}{\big(t\hat{a}(t)\big)^2} \left(\left(1 - \frac{\beta^2}{(t-\beta)^2}\right) \hat{a}(t) + \frac{t^2}{t-\beta} \tfrac{\mathrm{d}}{\mathrm{d}t}\hat{a}(t) + \frac{\beta^2}{(t-\beta)^2}\hat{a}(\beta) \right) \\ &= \big(\beta\hat{a}(\beta)\big)^2 \left(\frac{\hat{a}(t) + t\frac{\mathrm{d}}{\mathrm{d}t}\hat{a}(t)}{\big(t\hat{a}(t)\big)^2} + \beta \frac{\frac{\mathrm{d}}{\mathrm{d}t}\hat{a}(t)}{t(t-\beta)\big(\hat{a}(t)\big)^2} + \beta^2 \frac{\hat{a}(\beta) - \hat{a}(t)}{\big(t(t-\beta)\hat{a}(t)\big)^2} \right). \end{aligned} \tag{3.39}$$

Now we observe that

$$\frac{1}{t} < \frac{1}{t-\beta} < \frac{c'}{c'-\beta}\frac{1}{t}, \quad t > c' > \beta. \tag{3.40}$$

Moreover, Lemma 1.1.5(iv) implies that $t\hat{a}(t) \geq c'\hat{a}(c')$ for every $t > c'$ and

$$\frac{-\frac{\mathrm{d}}{\mathrm{d}t}\hat{a}(t)}{\hat{a}(t)} \leq \frac{1}{t}, \quad t > 0. \tag{3.41}$$

Using the above and Lemma 1.1.5(v) we obtain that for every $t > c' > \beta$,

$$\int_{c'}^{\infty} t^{1-2\eta} \frac{-\frac{\mathrm{d}}{\mathrm{d}t}\hat{a}(t)}{t(t-\beta)\big(\hat{a}(t)\big)^2}\, \mathrm{d}t \leq \frac{1}{(c'-\beta)\hat{a}(c')} \int_{c'}^{\infty} t^{-1-2\eta}\, \mathrm{d}t < \infty,$$

and

$$\int_{c'}^{\infty} t^{1-2\eta} \frac{\hat{a}(\beta) - \hat{a}(t)}{\big(t(t-\beta)\hat{a}(t)\big)^2}\, \mathrm{d}t \leq \frac{1}{\beta(c'-\beta)\hat{a}(c')} \int_{c'}^{\infty} t^{-1-2\eta}\, \mathrm{d}t < \infty.$$

Combined with (3.39) this implies that (3.38) holds if and only if

$$\int_{c'}^{\infty} t^{-1-2\eta} \frac{\hat{a}(t) + t\frac{\mathrm{d}}{\mathrm{d}t}\hat{a}(t)}{\big(\hat{a}(t)\big)^2}\, \mathrm{d}t < \infty \tag{3.42}$$

for some $c' > 1$ sufficiently large. Note that (3.42) holds with $c' > 1$ if and only if (3.37) holds. This shows that ξ belongs to $(H, \mathrm{D}(A^*))_{1-\eta,2}$ if and only if (3.37) holds and the proposition is proved. □

Proposition 3.5.3 *The restriction $J|_{\mathrm{D}(A)}$ is uniquely extensible to a bounded linear functional on $(H, \mathrm{D}(A))_{\eta,2}$ for every $\eta \in (0,1)$ such that $\eta > \frac{1-\alpha(a)}{2}$.*

PROOF: Let $\eta \in (0,1)$ be such that $\eta > \frac{1-\alpha(a)}{2}$. As $\mathrm{D}(A)$ is dense in $(H, \mathrm{D}(A))_{\eta,2}$ by Lemma 1.4.17 we have to show that there exists $M > 0$ such that

$$|J(\varphi)| \leq M \|\varphi\|_{(H,\mathrm{D}(A))_{\eta,2}}, \quad \varphi \in \mathrm{D}(A).$$

We observe that $(H, \mathrm{D}(A))_{\eta,2} = (H, \mathrm{D}(A - \beta I_H))_{\eta,2}$ by Lemma 1.4.15 and, since $0 \in \rho(A - \beta I_H)$ by Lemma 3.3.10, $(H, \mathrm{D}(A - \beta I_H))_{\eta,2} = \mathrm{D}((\beta I_H - A)^\eta)$ by Theorem 1.4.27. Moreover, the norms of $(H, \mathrm{D}(A))_{\eta,2}$, $(H, \mathrm{D}(A - \beta I_H))_{\eta,2}$, and the norm $\varphi \mapsto \|(\beta I_H - A)^\eta \varphi\|_H$ on $\mathrm{D}((\beta I_H - A)^\eta)$ are equivalent. Therefore it is sufficient to show that there exists $M > 0$ such that

$$|J(\varphi)| \leq M \big\|(\beta I_H - A)^\eta \varphi\big\|_H, \quad \varphi \in \mathrm{D}(A). \tag{3.43}$$

We shall make use of the function ξ given by Definition 3.5.1. By Proposition 3.5.2 we have that ξ belongs to $(H, \mathrm{D}(A^*))_{1-\eta,2}$. Since $(H, \mathrm{D}(A^*))_{1-\eta,2} = \mathrm{D}((\beta I_H - A^*)^{1-\eta})$ with the norm of $(H, \mathrm{D}(A^*))_{1-\eta,2}$ equivalent to the norm $\varphi \mapsto \|(\beta I_H - A^*)^{1-\eta}\varphi\|_H$ on $\mathrm{D}((\beta I_H - A^*)^{1-\eta})$, there exists $C > 0$ such that

$$\big\|(\beta I_H - A^*)^{1-\eta}\xi\big\|_H \leq C \, \|\xi\|_{(H,\mathrm{D}(A^*))_{1-\eta,2}}. \tag{3.44}$$

Using Propositions 3.3.7 and 1.4.26 and Lemma 1.4.28, we have

$$\begin{aligned}
J(\varphi) &= \big(J \circ (\beta I_H - A)^{-1} \circ (\beta I_H - A)\big)(\varphi) \\
&= \frac{1}{\beta \hat{a}(\beta)} \int_{[0,\infty)} \frac{\kappa}{\kappa + \beta} \big((\beta I_H - A)\varphi\big)(\kappa)\, \nu(\mathrm{d}\kappa) \\
&= \frac{1}{\beta \hat{a}(\beta)} \int_{[0,\infty)} \big((\beta I_H - A)\varphi\big)(\kappa)\, \xi(\kappa)(\kappa + \beta)\, \nu(\mathrm{d}\kappa) \\
&= \frac{1}{\beta \hat{a}(\beta)} \big\langle (\beta I_H - A)\varphi, \xi \big\rangle_H \\
&= \frac{1}{\beta \hat{a}(\beta)} \big\langle (\beta I_H - A)\varphi, (\beta I_H - A^*)^{\eta-1}(\beta I_H - A^*)^{1-\eta}\xi \big\rangle_H \\
&= \frac{1}{\beta \hat{a}(\beta)} \big\langle (\beta I_H - A)^\eta\varphi, (\beta I_H - A^*)^{1-\eta}\xi \big\rangle_H, \quad \varphi \in \mathrm{D}(A).
\end{aligned}$$

Hence, using (3.44) we obtain that

$$\begin{aligned} |J(\varphi)| &\le \frac{1}{\beta\hat{a}(\beta)} \big\|(\beta I_H - A)^\eta \varphi\big\|_H \big\|(\beta I_H - A^*)^{1-\eta}\xi\big\|_H \\ &\le \frac{C}{\beta\hat{a}(\beta)} \|(\beta I_H - A)^\eta \varphi\|_H \|\xi\|_{(H,\mathrm{D}(A^*))_{1-\eta,2}}, \qquad \varphi \in \mathrm{D}(A). \end{aligned}$$

This shows that (3.43) holds with $M := \frac{C}{\beta\hat{a}(\beta)}\|\xi\|_{(H,\mathrm{D}(A^*))_{1-\eta,2}}$. □

3.5.2 The function π

In this section we introduce a function $\pi : [0,\infty) \to \mathbb{R}$. One of the tasks of π is to solve the inhomogeneous Volterra equation in Section 3.7. We prove that π has two properties. The first is that π satisfies $\int_{[0,\infty)} (J(\pi) - \kappa\pi(\kappa))\,\nu(\mathrm{d}\kappa) = 1$. The second is that π belongs to an interpolation space $(H, \mathrm{D}(A))_{\theta,2}$ for some $\theta \in (0,1)$.

Definition 3.5.4 The function $\pi : [0,\infty) \to \mathbb{R}$ is defined by

$$\pi(\kappa) := \frac{1}{\beta\hat{a}(\beta)} \frac{1}{\kappa + \beta}, \quad \kappa \ge 0.$$

To prove that π has the first property we define a linear functional $I : \mathrm{D}(J) \to \mathbb{R}$ by

$$I(\varphi) := \int_{[0,\infty)} \big(J(\varphi) - \kappa\varphi(\kappa)\big)\,\nu(\mathrm{d}\kappa), \quad \varphi \in \mathrm{D}(J).$$

Note that I is well-defined by Lemma 3.3.5.

Lemma 3.5.5 *The function π belongs to $\mathrm{D}(J)$, and $J(\pi) = \frac{1}{\beta\hat{a}(\beta)}$ and $I(\pi) = 1$. Furthermore, $\varphi - I(\varphi)\pi$ belongs to $\mathrm{D}(A)$ for every $\varphi \in \mathrm{D}(J)$.*

PROOF: Since π belongs to H by Lemma 1.1.5(i), and

$$\frac{1}{\beta\hat{a}(\beta)} - \kappa\pi(\kappa) = \beta\pi(\kappa), \quad \kappa \ge 0,$$

this shows that π belongs to $\mathrm{D}(J)$ with $J(\pi) = \frac{1}{\beta\hat{a}(\beta)}$, and, using Lemma 1.1.5(i),

$$I(\pi) = \int_{[0,\infty)} \beta\pi(\kappa)\,\nu(\mathrm{d}\kappa) = 1.$$

Now the last part of the lemma follows from the observation that if φ belongs to $\mathrm{D}(J)$, then φ belongs to $\mathrm{D}(A)$ if and only if $I(\varphi) = 0$. □

To prove that π has the second property we use the following lemma.

Lemma 3.5.6 *It holds that*

$$J\big((\lambda I_H - A)^{-1}\pi\big) = \frac{1}{\lambda-\beta}\left(\frac{1}{\beta\hat{a}(\beta)} - \frac{1}{\lambda\hat{a}(\lambda)}\right), \quad \lambda > 0.$$

PROOF: Using Proposition 3.3.7 we have for every $\lambda > 0$,

$$\begin{aligned} J\big((\lambda I_H - A)^{-1}\pi\big) &= \frac{1}{\lambda\hat{a}(\lambda)}\int_{[0,\infty)} \frac{\kappa\pi(\kappa)}{\kappa+\lambda}\,\nu(\mathrm{d}\kappa) \\ &= \frac{1}{\beta\hat{a}(\beta)\lambda\hat{a}(\lambda)}\int_{[0,\infty)} \frac{\kappa}{(\kappa+\lambda)(\kappa+\beta)}\,\nu(\mathrm{d}\kappa) \\ &= \frac{1}{\beta\hat{a}(\beta)\lambda\hat{a}(\lambda)}\int_{[0,\infty)} \frac{1}{\lambda-\beta}\left(\frac{\lambda}{\kappa+\lambda} - \frac{\beta}{\kappa+\beta}\right)\nu(\mathrm{d}\kappa) \\ &= \frac{1}{\beta\hat{a}(\beta)\lambda(\lambda-\beta)\hat{a}(\lambda)}\big(\lambda\hat{a}(\lambda) - \beta\hat{a}(\beta)\big) = \frac{1}{\lambda-\beta}\left(\frac{1}{\beta\hat{a}(\beta)} - \frac{1}{\lambda\hat{a}(\lambda)}\right). \end{aligned}$$

□

Proposition 3.5.7 *The function π belongs to $(H, \mathrm{D}(A))_{\theta,2}$ for every $\theta \in (0,1)$ such that $\theta < \frac{1+\alpha(a)}{2}$.*

PROOF: Let $\theta \in (0,1)$. We show that π belongs to $(H, \mathrm{D}(A))_{\theta,2}$ if and only if

$$\int_1^\infty t^{2\theta-3}\frac{1}{(\hat{a}(t))^2}\big(\hat{a}(t) + t\tfrac{\mathrm{d}}{\mathrm{d}t}\hat{a}(t)\big)\,\mathrm{d}t < \infty. \tag{3.45}$$

This would prove the proposition. Indeed, if $\theta < \frac{1+\alpha(a)}{2}$, then $2\theta - 1 < \alpha(a)$, so that (3.45) holds by definition of $\alpha(a)$ and hence, π belongs to $(H, \mathrm{D}(A))_{\theta,2}$. By definition of interpolation spaces we have that π belongs to $(H, \mathrm{D}(A))_{\theta,2}$ if

$$\int_c^\infty t^{2\theta-1}\big\|A(tI_H - A)^{-1}\pi\big\|_H^2\,\mathrm{d}t < \infty \tag{3.46}$$

for some $c > 1$ sufficiently large. By Proposition 3.3.7 and Lemma 3.5.6 we have for all $t > 0$ and $\kappa \geq 0$,

$$\begin{aligned} \big(A(tI_H - A)^{-1}\pi\big)(\kappa) &= -\pi(\kappa) + t\frac{\pi(\kappa) + J\big((tI_H - A)^{-1}\pi\big)}{\kappa+t} \\ &= J\big((tI_H - A)^{-1}\pi\big)\frac{t}{\kappa+t} - \frac{1}{\beta\hat{a}(\beta)}\frac{\kappa}{(\kappa+t)(\kappa+\beta)} \\ &= \frac{1}{\hat{a}(\beta)}\frac{1}{t-\beta}\frac{1}{\kappa+\beta} - \frac{1}{(t-\beta)\hat{a}(t)}\frac{1}{\kappa+t}. \end{aligned}$$

Using Lemma 1.1.5(i), (ii), and (iii) we obtain that for every $t > 0$,

$$\begin{aligned}
&\left\|A(tI_H - A)^{-1}\pi\right\|_H^2 \qquad (3.47)\\
&= \frac{1}{\hat{a}(\beta)(t-\beta)^2} - \frac{2}{\hat{a}(\beta)(t-\beta)^2} + \frac{\hat{a}(t) + t\frac{\mathrm{d}}{\mathrm{d}t}\hat{a}(t) - \beta\frac{\mathrm{d}}{\mathrm{d}t}\hat{a}(t)}{\big((t-\beta)\hat{a}(t)\big)^2}\\
&= -\frac{1}{\hat{a}(\beta)(t-\beta)^2} + \frac{\hat{a}(t) + t\frac{\mathrm{d}}{\mathrm{d}t}\hat{a}(t)}{\big((t-\beta)\hat{a}(t)\big)^2} - \beta\frac{\frac{\mathrm{d}}{\mathrm{d}t}\hat{a}(t)}{\big((t-\beta)\hat{a}(t)\big)^2}.
\end{aligned}$$

By (3.40), (3.41), and the fact that $t\hat{a}(t) \geq c\hat{a}(c)$ for every $t > c$ by Lemma 1.1.5(iv), we observe that

$$\int_c^\infty t^{2\theta-1}\frac{1}{(t-\beta)^2}\,\mathrm{d}t \leq \left(\frac{c}{c-\beta}\right)^2 \int_c^\infty t^{2\theta-3}\,\mathrm{d}t < \infty,$$

and

$$\int_c^\infty t^{2\theta-1}\frac{-\frac{\mathrm{d}}{\mathrm{d}t}\hat{a}(t)}{\big((t-\beta)\hat{a}(t)\big)^2}\,\mathrm{d}t \leq \frac{c}{(c-\beta)^2\hat{a}(c)}\int_c^\infty t^{2\theta-3}\,\mathrm{d}t < \infty.$$

Combined with (3.47) this implies that (3.46) holds if and only if

$$\int_c^\infty t^{2\theta-1}\frac{\hat{a}(t) + t\frac{\mathrm{d}}{\mathrm{d}t}\hat{a}(t)}{\big((t-\beta)\hat{a}(t)\big)^2}\,\mathrm{d}t < \infty. \qquad (3.48)$$

Note that (3.48) holds with $c > 1$ if and only if (3.45) holds. This shows that π belongs to $(H, \mathrm{D}(A))_{\theta,2}$ if and only if (3.45) holds and the proposition is proved. □

The next corollary is an immediate result of Lemma 3.5.5 and Proposition 3.5.7.

Corollary 3.5.8 *It holds that* $\mathrm{D}(J) = \mathrm{D}(A) \oplus \mathrm{span}\{\pi\}$. *Even more,* $\mathrm{D}(J) \subseteq (H, \mathrm{D}(A))_{\theta,2}$ *for every* $\theta \in (0,1)$ *such that* $\theta < \frac{1+\alpha(a)}{2}$.

3.6 The homogeneous problem II

In Section 3.4 we have considered the homogeneous Volterra equation

$$\begin{aligned}
\frac{\mathrm{d}}{\mathrm{d}t}\int_{-\infty}^t a(t-s)u(s)\,\mathrm{d}s &= 0, \qquad t > 0,\\
u(t) &= u_0(t), \quad t \leq 0,
\end{aligned} \qquad (3.49)$$

where the kernel $a : (0,\infty) \to \mathbb{R}$ belongs to the class $\mathcal{K}$ and where the function $u_0 : (-\infty, 0] \to \mathbb{R}$ satisfies Hypothesis 3.4.1. With the results of Section 3.5 we can add a regularity result to Theorem 3.4.8.

Proposition 3.6.1 *If u_0 satisfies Hypothesis 3.4.1, then the unique solution to problem (3.49) belongs to $C^{0,1-\eta}[0,T]$ for all $T > 0$ and $\eta \in (0,1)$ such that $\eta > \frac{1-\alpha(a)}{2}$.*

PROOF: Let u be the solution to problem (3.49) and let $T > 0$ and $\eta \in (0,1)$ such that $\eta > \frac{1-\alpha(a)}{2}$. By Theorem 3.4.8 we have $u(t) = J(S(t)\psi_0)$ for every $t \geq 0$, where ψ_0 belongs to $\mathrm{D}(A)$ by Lemma 3.4.4. Since $A - \beta I_H$ is the infinitesimal generator of an analytic semigroup on H with $0 \in \rho(A - \beta I_H)$ by Proposition 1.4.13 and Lemma 3.3.10, and since $(H, \mathrm{D}(A))_{\eta,2} = (H, \mathrm{D}(A-\beta I_H))_{\eta,2}$ by Lemma 1.4.15, it follows from Proposition 1.4.31(ii) that the mapping $t \mapsto S(t)\psi_0$ belongs to $C^{0,1-\eta}([0,T]; (H, \mathrm{D}(A))_{\eta,2})$. Now the continuity of J on $(H, \mathrm{D}(A))_{\eta,2}$ by Proposition 3.5.3 implies that u belongs to $C^{0,1-\eta}[0,T]$. □

In this section we solve problem (3.49) under weaker assumptions on u_0, assuming that u_0 only satisfies Hypothesis 3.4.1(i) and (ii), that is, u_0 satisfies the next hypothesis:

Hypothesis 3.6.2 The function $u_0 : (-\infty, 0] \to \mathbb{R}$ is Borel measurable and has the following properties:

(i) There exist $M_1 > 0$ and $\omega > 0$ such that $|u_0(t)| \leq M_1 \mathrm{e}^{\omega t}$ for every $t \leq 0$;

(ii) There exist $M_2 > 0$ and $\delta > 0$ such that $|u_0(0) - u_0(t)| \leq M_2|t|$ for every $t \in [-\delta, 0]$.

We start with a special choice of u_0 for which we recall that the resolvent of the first kind of a is a function $b \in \mathcal{K}$ such that

$$\int_0^t a(t-s)b(s)\,\mathrm{d}s = 1, \quad t > 0. \tag{3.50}$$

Proposition 3.6.3 *Let $u_0 : \mathbb{R} \to \mathbb{R}$ be given by $u_0(t) := \mathrm{e}^{\alpha t}$ for every $t \in \mathbb{R}$ where $\alpha > 0$. Then problem (3.49) admits a unique solution u, and u has the representation*

$$u(t) = \begin{cases} u_0(t) - \alpha\hat{a}(\alpha)\int_0^t b(t-s)u_0(s)\,\mathrm{d}s, & t > 0, \\ u_0(t), & t \leq 0, \end{cases} \tag{3.51}$$

where b is the resolvent of the first kind of a.

PROOF: Note that u given by (3.51) is continuous. Using Lemma 1.1.5 we have for every $t \geq 0$,

$$\begin{aligned}\int_{-\infty}^t a(t-s)|u(s)|\,\mathrm{d}s &= \int_{-\infty}^0 a(t-s)\mathrm{e}^{\alpha s}\,\mathrm{d}s + \int_0^t a(t-s)|u(s)|\,\mathrm{d}s \\ &\leq \hat{a}(\alpha)\mathrm{e}^{\alpha t} + \max\{|u(s)|;\, s \in [0,t]\}\int_0^t a(s)\,\mathrm{d}s < \infty.\end{aligned}$$

Thus u satisfies Definition 3.1.1(i). Now we use the associative property of convolutions and (3.50) to obtain

$$\begin{aligned}
&\int_{-\infty}^{t} a(t-s)u(s)\,\mathrm{d}s \\
&= \int_{-\infty}^{t} a(t-s)u_0(s)\,\mathrm{d}s - \alpha\hat{a}(\alpha)\int_0^t a(t-s)\left(\int_0^s b(s-\sigma)u_0(\sigma)\,\mathrm{d}\sigma\right)\mathrm{d}s \\
&= \hat{a}(\alpha)\mathrm{e}^{\alpha t} - \alpha\hat{a}(\alpha)\int_0^t\left(\int_0^{t-s} a(t-s-\sigma)b(\sigma)\,\mathrm{d}\sigma\right)u_0(s)\,\mathrm{d}s \\
&= \hat{a}(\alpha)u_0(t) - \alpha\hat{a}(\alpha)\int_0^t u_0(s)\,\mathrm{d}s, \quad t \geq 0.
\end{aligned}$$

This implies that u satisfies Definition 3.1.1(ii). Even more, u satisfies Definition 3.1.1(iii), since

$$\frac{\mathrm{d}}{\mathrm{d}t}\int_{-\infty}^{t} a(t-s)u(s)\,\mathrm{d}s = \hat{a}(\alpha)\tfrac{\mathrm{d}}{\mathrm{d}t}u_0(t) - \alpha\hat{a}(\alpha)u_0(t) = 0, \quad t > 0.$$

□

Theorem 3.6.4 *Let u_0 satisfy Hypothesis 3.6.2. If a belongs to $\mathcal{K}$ with $\alpha(a) \in (0,1)$, then problem (3.49) admits a unique solution u. Moreover, u is real analytic in $(0,\infty)$, belongs to $C^{0,\zeta}[0,T]$ for all $T > 0$ and $\zeta \in [0,\alpha(a))$, and has the representation*

$$u(t) = \begin{cases} J\big(S(t)\psi_0\big), & t > 0, \\ u_0(t), & t \leq 0, \end{cases}$$

where ψ_0 is given by

$$\psi_0(\kappa) := \int_0^\infty \mathrm{e}^{-\kappa s}u_0(-s)\,\mathrm{d}s, \quad \kappa \geq 0.$$

PROOF: Let $T > 0$ and $\zeta \in [0,\alpha(a))$, and let $\eta := \frac{1-\zeta}{2}$ and $\theta := \frac{1+\zeta}{2}$. Then we have $\theta - \eta = \zeta$ and $0 < \frac{1-\alpha(a)}{2} < \eta \leq \theta < \frac{1+\alpha(a)}{2} < 1$. Let the functions $u_1, u_2 : (-\infty,0] \to \mathbb{R}$ and $\psi_1, \psi_2 : [0,\infty) \to \mathbb{R}$ be defined by respectively $u_1(t) := u_0(t) - \gamma\mathrm{e}^{\beta t}$ and $u_2(t) := \gamma\mathrm{e}^{\beta t}$ for every $t \leq 0$, and

$$\psi_j(\kappa) := \int_0^\infty \mathrm{e}^{-\kappa s}u_j(-s)\,\mathrm{d}s, \quad \kappa \geq 0,\ j = 1,2,$$

where

$$\gamma := \frac{1}{\beta\hat{a}(\beta)}\left(u_0(0)\,a(+\infty) + \int_0^\infty \big(-\tfrac{\mathrm{d}}{\mathrm{d}s}a(s)\big)\big(u_0(0) - u_0(-s)\big)\,\mathrm{d}s\right).$$

Note that $u_0 = u_1 + u_2$ and $\psi_0 = \psi_1 + \psi_2$, and that u_1 and u_2 satisfy Hypothesis 3.6.2. Even more, u_1 also satisfies Hypothesis 3.4.1(iii). Indeed, this follows from Lemma 3.4.2 since we have, using the definition of γ and Lemma 1.1.6(ii),

$$u_1(0)\,a(+\infty) + \int_0^\infty \big(-\tfrac{\mathrm{d}}{\mathrm{d}s}a(s)\big)\big(u_1(0) - u_1(-s)\big)\,\mathrm{d}s$$
$$= u_0(0)\,a(+\infty) + \int_0^\infty \big(-\tfrac{\mathrm{d}}{\mathrm{d}s}a(s)\big)\big(u_0(0) - u_0(-s)\big)\,\mathrm{d}s -$$
$$\gamma\left(a(+\infty) + \int_0^\infty \big(-\tfrac{\mathrm{d}}{\mathrm{d}s}a(s)\big)\big(1 - \mathrm{e}^{-\beta s}\big)\,\mathrm{d}s\right)$$
$$= \beta\hat{a}(\beta)\gamma - \gamma\left(\beta \lim_{s\downarrow 0} s a(s)\frac{1-\mathrm{e}^{-\beta s}}{\beta s} + \beta\hat{a}(\beta)\right) = 0.$$

Therefore Theorem 3.4.8 implies that the function $\tilde{u}_1 : \mathbb{R} \to \mathbb{R}$ defined by

$$\tilde{u}_1(t) = \begin{cases} J\big(S(t)\psi_1\big), & t > 0, \\ u_1(t), & t \leq 0, \end{cases}$$

is the unique solution to problem (3.49) with u_0 replaced by u_1, and $\tilde{u}_1$ is real analytic in $(0,\infty)$. Furthermore, it is a result of Proposition 3.6.1 that $\tilde{u}_1$ belongs to $C^{0,1-\eta}[0,T]$.

Now we extend u_2 to a function defined on $\mathbb{R}$ by $u_2(t) := \gamma\mathrm{e}^{\beta t}$ for every $t \in \mathbb{R}$. Then it is a result of Proposition 3.6.3 that the function $\tilde{u}_2 : \mathbb{R} \to \mathbb{R}$ defined by

$$\tilde{u}_2(t) := \begin{cases} u_2(t) - \beta\hat{a}(\beta)\int_0^t b(t-s)u_2(s)\,\mathrm{d}s, & t > 0, \\ u_2(t), & t \leq 0, \end{cases}$$

where b denotes the resolvent of the first kind of a, is the unique solution to problem (3.49) with u_0 replaced by u_2. By means of the Laplace transform we shall prove that

$$\tilde{u}_2(t) = \begin{cases} J\big(S(t)\psi_2\big), & t > 0, \\ u_2(t), & t \leq 0. \end{cases} \tag{3.52}$$

We observe that $\hat{u}_2(\lambda) = \frac{\gamma}{\lambda-\beta}$ for every $\lambda > \beta$ and $\hat{a}(\lambda)\hat{b}(\lambda) = \frac{1}{\lambda}$ for every $\lambda > 0$, the latter by (3.50) and the fact that the Laplace transform of a convolution is the product of the individual Laplace transforms. Moreover, recalling that π is given by

$$\pi(\kappa) := \frac{1}{\beta\hat{a}(\beta)}\frac{1}{\kappa+\beta}, \quad \kappa \geq 0,$$

we have

$$\psi_2(\kappa) = \int_0^\infty \mathrm{e}^{-\kappa s}u_2(-s)\,\mathrm{d}s = \frac{\gamma}{\kappa+\beta} = \gamma\beta\hat{a}(\beta)\pi(\kappa), \quad \kappa \geq 0.$$

We remark that π, and thus ψ_2, belongs to $(H,\mathrm{D}(A))_{\theta,2}$ by Proposition 3.5.7, and that $(H,\mathrm{D}(A))_{\theta,2} \subseteq (H,\mathrm{D}(A))_{\eta,2}$ by Lemma 1.4.16. Therefore $\int_0^\infty \mathrm{e}^{-\lambda t} S(t)\pi\,\mathrm{d}t$ exists as a Bochner integral in $(H,\mathrm{D}(A))_{\eta,2}$ whenever $\lambda > \beta$. Indeed, since $A-\beta I_H$ is the infinitesimal generator of the analytic semigroup $\{\mathrm{e}^{-\beta t}S(t)\}_{t\geq 0}$ on H with $0 \in \rho(A-\beta I_H)$ by Propositions 1.4.9 and 1.4.13 and Lemma 3.3.10, and since $(H,\mathrm{D}(A))_{\eta,2} = (H,\mathrm{D}(A-\beta I_H))_{\eta,2}$ by Lemma 1.4.15, we have by Corollary 1.4.30(i) that there exists $M>0$ such that

$$\int_0^\infty \big\|\mathrm{e}^{-\lambda t} S(t)\pi\big\|_{(H,\mathrm{D}(A))_{\eta,2}}\,\mathrm{d}t \leq \|\pi\|_{(H,\mathrm{D}(A))_{\eta,2}} \int_0^\infty \mathrm{e}^{-\lambda t}\|S(t)\|_{\mathcal{L}((H,\mathrm{D}(A))_{\eta,2})}\,\mathrm{d}t$$
$$\leq M\|\pi\|_{(H,\mathrm{D}(A))_{\eta,2}}\frac{1}{\lambda-\beta} < \infty, \qquad \lambda>\beta.$$

Now we use the above and Lemma 3.5.6, Proposition 1.4.6, and the continuity of J on $(H,\mathrm{D}(A))_{\eta,2}$ by Proposition 3.5.3, to obtain

$$\int_0^\infty \mathrm{e}^{-\lambda t}\tilde{u}_2(t)\,\mathrm{d}t = \int_0^\infty \mathrm{e}^{-\lambda t}\left(u_2(t) - \beta\hat{a}(\beta)\int_0^t b(t-s)u_2(s)\,\mathrm{d}s\right)\mathrm{d}t$$
$$= \hat{u}_2(\lambda) - \beta\hat{a}(\beta)\hat{b}(\lambda)\hat{u}_2(\lambda) = \gamma\beta\hat{a}(\beta)\frac{1}{\lambda-\beta}\left(\frac{1}{\beta\hat{a}(\beta)} - \frac{1}{\lambda\hat{a}(\lambda)}\right)$$
$$= \gamma\beta\hat{a}(\beta)J\big((\lambda I_H - A)^{-1}\pi\big) = \gamma\beta\hat{a}(\beta)J\left(\int_0^\infty \mathrm{e}^{-\lambda t}S(t)\pi\,\mathrm{d}t\right)$$
$$= \gamma\beta\hat{a}(\beta)\int_0^\infty \mathrm{e}^{-\lambda t}J\big(S(t)\pi\big)\,\mathrm{d}t = \int_0^\infty \mathrm{e}^{-\lambda t}J\big(S(t)\psi_2\big)\,\mathrm{d}t, \qquad \lambda>\beta.$$

This shows that the Laplace transforms of $\tilde{u}_2$ and the mapping $t\mapsto J(S(t)\psi_2)$ are equal on (β,∞). Hence, (3.52) holds as a consequence of the inversion theorem for Laplace transforms. We observe that $\tilde{u}_2$ is real analytic in $(0,\infty)$ by the analyticity of $\{S(t)\}_{t\geq 0}$, Proposition 1.4.29(i), and the continuity of J on $(H,\mathrm{D}(A))_{\eta,2}$. Moreover, Proposition 1.4.31(iii) implies that ψ_2 belongs to $C^{0,\theta-\eta}\big([0,T];(H,\mathrm{D}(A))_{\eta,2}\big)$ and hence, it follows from the continuity of J on $(H,\mathrm{D}(A))_{\eta,2}$ that $\tilde{u}_2$ belongs to $C^{0,\theta-\eta}[0,T]$. Finally, the theorem is proved by defining $u := \tilde{u}_1 + \tilde{u}_2$. □

Corollary 3.6.5 *Let u be the unique solution to problem (3.49) under the conditions stated in Theorem 3.6.4. Then the following holds:*

$$\big(S(t)\psi_0\big)(\kappa) = \int_0^\infty \mathrm{e}^{-\kappa s}u(t-s)\,\mathrm{d}s, \quad t\geq 0,\ \kappa\geq 0.$$

PROOF: We shall use the Laplace transform. Let $\theta,\eta \in (0,1)$ be such that $0 < \frac{1-\alpha(a)}{2} < \eta \leq \theta < \frac{1+\alpha(a)}{2} < 1$. First we observe that ψ_0 belongs to $\mathrm{D}(J)$

by Lemma 3.4.4 and hence, to $(H, \mathrm{D}(A))_{\theta,2}$ by Corollary 3.5.8. By the same arguments as in the proof of Theorem 3.6.4 we have that $\int_0^\infty \mathrm{e}^{-\lambda t} S(t)\psi_0 \,\mathrm{d}t$ exists as a Bochner integral in $(H, \mathrm{D}(A))_{\eta,2}$ for every $\lambda > \beta$. Using the representation of u, the continuity of J on $(H, \mathrm{D}(A))_{\eta,2}$ by Proposition 3.5.3, and Proposition 1.4.6, it follows that for every $\lambda > \beta$,

$$\hat{u}(\lambda) = \int_0^\infty \mathrm{e}^{-\lambda t} J\big(S(t)\psi_0\big)\,\mathrm{d}t = J\left(\int_0^\infty \mathrm{e}^{-\lambda t} S(t)\psi_0 \,\mathrm{d}t\right) = J\big((\lambda I_H - A)^{-1}\psi_0\big).$$

Now we use Propositions 1.4.6 and 3.3.7, the fact that the Laplace transform of a convolution is the product of the individual Laplace transforms, and the definition of ψ_0 to obtain

$$\begin{aligned}
&\int_0^\infty \mathrm{e}^{-\lambda t}\big(S(t)\psi_0\big)(\kappa)\,\mathrm{d}t = \big((\lambda I_H - A)^{-1}\psi_0\big)(\kappa) \\
&= \frac{\psi_0(\kappa) + J\big((\lambda I_H - A)^{-1}\psi_0\big)}{\kappa+\lambda} = \frac{\psi_0(\kappa)}{\kappa+\lambda} + \frac{\hat{u}(\lambda)}{\kappa+\lambda} \\
&= \int_0^\infty \mathrm{e}^{-\lambda t}\left(\mathrm{e}^{-\kappa t}\psi_0(\kappa) + \int_0^t \mathrm{e}^{-\kappa(t-s)}u(s)\,\mathrm{d}s\right)\mathrm{d}t \\
&= \int_0^\infty \mathrm{e}^{-\lambda t}\left(\mathrm{e}^{-\kappa t}\int_0^\infty \mathrm{e}^{-\kappa s}u_0(-s)\,\mathrm{d}s + \int_0^t \mathrm{e}^{-\kappa(t-s)}u(s)\,\mathrm{d}s\right)\mathrm{d}t \\
&= \int_0^\infty \mathrm{e}^{-\lambda t}\left(\int_0^\infty \mathrm{e}^{-\kappa s}u(t-s)\,\mathrm{d}s\right)\mathrm{d}t, \quad \lambda > \beta,\ \kappa \geq 0.
\end{aligned}$$

Therefore the corollary is a consequence of the inversion theorem for Laplace transforms. □

3.7 The inhomogeneous problem

We consider the inhomogeneous Volterra equation

$$\begin{aligned}
\frac{\mathrm{d}}{\mathrm{d}t}\int_{-\infty}^t a(t-s)u(s)\,\mathrm{d}s &= f(t), \quad t > 0, \\
u(t) &= u_0(t), \quad t \leq 0,
\end{aligned} \tag{3.53}$$

where the kernel $a : (0,\infty) \to \mathbb{R}$ belongs to the class $\mathcal{K}$ and where the function $f : [0,\infty) \to \mathbb{R}$ belongs to $L^1(0,T)$ for every $T > 0$. Without loss of generality we assume that u_0 is identically zero.

To be able to study existence and regularity of solutions to problem (3.53) we call to mind our analytic semigroup approach as explained in Section 3.1. We recall that the function $\psi : [0,\infty) \times [0,\infty) \to \mathbb{R}$ is defined by

$$\psi(t,\kappa) := \int_{-\infty}^{t} \mathrm{e}^{-\kappa(t-s)} u(s)\,\mathrm{d}s, \quad t \geq 0,\, \kappa \geq 0,$$

where the function $u : \mathbb{R} \to \mathbb{R}$ is such that (3.53) holds. Also we recall that ψ satisfies the initial value problem

$$\begin{aligned} \tfrac{\partial}{\partial t}\psi(t,\kappa) &= u(t) - \kappa\psi(t,\kappa), \quad t > 0,\, \kappa \geq 0, \\ \psi(0,\kappa) &= \psi_0(\kappa), \qquad\qquad\qquad\quad \kappa \geq 0, \end{aligned} \tag{3.54}$$

where the initial value $\psi_0 : [0,\infty) \to \mathbb{R}$ is defined by

$$\psi_0(\kappa) := \int_0^{\infty} \mathrm{e}^{-\kappa s} u_0(-s)\,\mathrm{d}s, \quad \kappa \geq 0,$$

and remember that ψ is subject to the constraint

$$\int_{[0,\infty)} \big(u(t) - \kappa\psi(t,\kappa)\big)\,\nu(\mathrm{d}\kappa) = f(t), \quad t > 0. \tag{3.55}$$

We have constructed a Hilbert space H, a linear functional J, and a linear operator A such that, heuristically, $\psi(t,\cdot)$ belongs to $\mathrm{D}(J)$ with $J(\psi(t,\cdot)) = u(t)$ and, in case f is identically zero, $\psi(t,\cdot)$ belongs to $\mathrm{D}(A)$ for every $t > 0$. In this case the initial value problem (3.54) together with the constraint (3.55) rewrites into the homogeneous abstract Cauchy problem

$$\begin{aligned} \tfrac{\mathrm{d}}{\mathrm{d}t}\varphi(t) &= A\varphi(t), \quad t > 0, \\ \varphi(0) &= \psi_0. \end{aligned} \tag{3.56}$$

However, in case f is not identically zero, (3.55) implies that $\psi(t,\cdot)$ does not belong to $\mathrm{D}(A)$ for every $t > 0$. Therefore the homogeneous abstract Cauchy problem (3.56) is of no use as a semigroup setting for the inhomogeneous Volterra equation (3.53).

The key to find a useful abstract Cauchy problem is the function π that was defined in Section 3.5.2 by

$$\pi(\kappa) := \frac{1}{\beta\hat{a}(\beta)} \frac{1}{\kappa + \beta}, \quad \kappa \geq 0.$$

To see why π is the key we define the linear operator $\bar{A} : \mathrm{D}(J) \to H$ by

$$(\bar{A}\varphi)(\kappa) := J(\varphi) - \kappa\varphi(\kappa), \quad \varphi \in \mathrm{D}(J),\, \kappa \geq 0,$$

and recall that the linear functional $I : \mathrm{D}(J) \to \mathbb{R}$ is given by

$$I(\varphi) := \int_{[0,\infty)} \big(J(\varphi) - \kappa\varphi(\kappa)\big)\, \nu(\mathrm{d}\kappa), \quad \varphi \in \mathrm{D}(J).$$

Note that Lemma 3.5.5 implies that $\bar{A}\pi = \beta\pi$ and that $\varphi - I(\varphi)\pi$ belongs to $\mathrm{D}(A)$ for every $\varphi \in \mathrm{D}(J)$. Therefore the semigroup approach suggests that, heuristically, $I(\psi(t,\cdot)) = f(t)$ for every $t > 0$ and

$$\begin{aligned} &\tfrac{\mathrm{d}}{\mathrm{d}t}\psi(t,\cdot) = \bar{A}\psi(t,\cdot) = A\big(\psi(t,\cdot) - f(t)\pi\big) + \beta f(t)\pi, \quad t > 0, \\ &\psi(0,\cdot) = \psi_0. \end{aligned} \tag{3.57}$$

Now we can rewrite the initial value problem (3.57) into an inhomogeneous abstract Cauchy problem

$$\begin{aligned} &\tfrac{\mathrm{d}}{\mathrm{d}t}\varphi(t) = A_{-1}\varphi(t) + f(t)(\beta I_H - A_{-1})\pi, \quad t > 0, \\ &\varphi(0) = \psi_0, \end{aligned} \tag{3.58}$$

in an extrapolation space H_{-1} of H. We choose the Hilbert space H_{-1} as the completion of H with respect to the norm $\|\cdot\|_{H_{-1}}$ given by

$$\|\varphi\|_{H_{-1}} := \big\|(\beta I_H - A)^{-1}\varphi\big\|_H, \quad \varphi \in H,$$

and let $A_{-1} : H \subseteq H_{-1} \to H_{-1}$ be the linear extension of A such that A_{-1} is the infinitesimal generator of the analytic semigroup $\{S_{-1}(t)\}_{t\geq 0}$ on H_{-1}, which denotes the continuous extension of $\{S(t)\}_{t\geq 0}$.

In Section 3.7.1 we study the inhomogeneous abstract Cauchy problem (3.58), to return to problem (3.53) in Section 3.7.2. We remark that we take u_0, and thus ψ_0, identically zero.

3.7.1 The inhomogeneous abstract Cauchy problem

In this section we consider the inhomogeneous abstract Cauchy problem

$$\begin{aligned} &\tfrac{\mathrm{d}}{\mathrm{d}t}\psi(t) = A_{-1}\psi(t) + f(t)(\beta I_H - A_{-1})\pi, \quad t > 0, \\ &\psi(0) = 0, \end{aligned} \tag{3.59}$$

where $f : [0,\infty) \to \mathbb{R}$ belongs to $L^1(0,T)$ for every $T > 0$. We use the following notion of solution:

Definition 3.7.1 The *mild solution* to problem (3.59) is the continuous function $\psi : [0,\infty) \to H_{-1}$ defined by

$$\psi(t) := \int_0^t f(s)(\beta I_H - A)S(t-s)\pi \, \mathrm{d}s, \quad t \geq 0,$$

where the integral is Bochner in H_{-1}.

With a view to apply the linear functional J to the mild solution ψ we investigate when ψ has values in an interpolation space $(H, \mathrm{D}(A))_{\eta,2}$ for some $\eta \in (0,1)$, along with some regularity properties. Then we state a result concerning the Laplace transform of ψ that will be useful in the next section.

Proposition 3.7.2 *The mild solution ψ has values in $(H, \mathrm{D}(A))_{\theta,2}$ for every $\theta \in (0,1)$ such that $\theta < \frac{1+\alpha(a)}{2}$.*

PROOF: Let $\theta, \eta \in (0,1)$ be such that $\eta < \theta < \frac{1+\alpha(a)}{2}$. Then π belongs to $(H, \mathrm{D}(A))_{\theta,2}$ by Proposition 3.5.7. We show that $\int_0^t f(s)(\beta I_H - A)S(t-s)\pi \,\mathrm{d}s$ exists as a Bochner integral in $(H, \mathrm{D}(A))_{\eta,2}$ for every $t > 0$. Since $A - \beta I_H$ is the infinitesimal generator of the analytic semigroup $\{\mathrm{e}^{-\beta t} S(t)\}_{t \geq 0}$ on H with $0 \in \rho(A - \beta I_H)$ by Propositions 1.4.9 and 1.4.13 and Lemma 3.3.10, and since $(H, \mathrm{D}(A))_{\eta,2} = (H, \mathrm{D}(A - \beta I_H))_{\eta,2}$ by Lemma 1.4.15, we have by Corollary 1.4.30(ii) that there exists $M > 0$ such that

$$\left\|(A - \beta I_H)\mathrm{e}^{-\beta t} S(t)\right\|_{\mathcal{L}((H,\mathrm{D}(A))_{\theta,2},(H,\mathrm{D}(A))_{\eta,2})} \leq M t^{-1+\theta-\eta}, \quad t > 0.$$

Therefore we have, using the function $g_{\theta-\eta}$ given by (1.2) in Example 1.3.3,

$$\begin{aligned}
&\int_0^t \left\|f(s)(\beta I_H - A)S(t-s)\pi\right\|_{(H,\mathrm{D}(A))_{\eta,2}} \mathrm{d}s \\
&\leq M\|\pi\|_{(H,\mathrm{D}(A))_{\theta,2}} \int_0^t \mathrm{e}^{\beta(t-s)}(t-s)^{-1+\theta-\eta}|f(s)| \,\mathrm{d}s \\
&\leq M\Gamma(\theta-\eta)\|\pi\|_{(H,\mathrm{D}(A))_{\theta,2}} \mathrm{e}^{\beta t} \int_0^t g_{\theta-\eta}(t-s)|f(s)| \,\mathrm{d}s < \infty, \qquad t > 0.
\end{aligned}$$

□

Proposition 3.7.3 *Let $f : [0,\infty) \to \mathbb{R}$ belong to $L^p(0,T)$ for some $p \in [1,\infty)$ and every $T > 0$. Let ψ be the mild solution to problem (3.59). Then for all $T > 0$ and $\theta, \eta \in (0,1)$ such that $\eta < \theta < \frac{1+\alpha(a)}{2}$ the following holds:*

(i) *If $p = 1$, then ψ belongs to $L^q\big(0,T;(H,\mathrm{D}(A))_{\eta,2}\big)$ for every $q \in \big[1, \frac{1}{1-(\theta-\eta)}\big)$;*

(ii) *If $p \in \big(1, \frac{1}{\theta-\eta}\big)$, then ψ belongs to $L^q\big(0,T;(H,\mathrm{D}(A))_{\eta,2}\big)$ where $q := \frac{p}{1-(\theta-\eta)p}$;*

(iii) *If $p = \frac{1}{\theta-\eta}$, then ψ belongs to $L^q\big(0,T;(H,\mathrm{D}(A))_{\eta,2}\big)$ for every $q \in [1,\infty)$;*

(iv) *If $p \in \big(\frac{1}{\theta-\eta}, \infty\big)$, then ψ belongs to $C_0^{0,\theta-\eta-\frac{1}{p}}\big([0,T];(H,\mathrm{D}(A))_{\eta,2}\big)$.*

In particular, if f belongs to $C_0^{0,\gamma}[0,T]$ for some $\gamma \in (0,1)$ and every $T>0$, then ψ belongs to

$$C^{[\gamma+\theta-\eta],\gamma+\theta-\eta-[\gamma+\theta-\eta]}\big([0,T];(H,\mathrm{D}(A))_{\eta,2}\big)$$

for all $T>0$ and $\theta,\eta \in (0,1)$ such that $\eta<\theta<\frac{1+\alpha(a)}{2}$ and $\theta-\eta \neq 1-\gamma$.

PROOF: The first part of the proposition follows from Proposition 1.4.31(iv)-(vii), since π belongs to $(H,\mathrm{D}(A))_{\theta,2}$ for every $\theta \in (0,1)$ such that $\theta < \frac{1+\alpha(a)}{2}$ by Proposition 3.5.7, since $A-\beta I_H$ is the infinitesimal generator of the analytic semigroup $\{\mathrm{e}^{-\beta t}S(t)\}_{t\geq 0}$ on H with $0 \in \rho(A-\beta I_H)$ by Propositions 1.4.9 and 1.4.13 and Lemma 3.3.10, and since $(H,\mathrm{D}(A))_{\eta,2} = (H,\mathrm{D}(A-\beta I_H))_{\eta,2}$ for every $\eta \in (0,1)$ by Lemma 1.4.15.
For the second part we fix $T>0$ and $\theta,\eta \in (0,1)$ such that $\eta<\theta<\frac{1+\alpha(a)}{2}$ and $\theta-\eta\neq 1-\gamma$. Then we have that the mapping $t \mapsto f(t)(\beta I_H - A_{-1})\pi$ belongs to $C_0^{0,\gamma}\big([0,T];(H_{-1},H)_{\theta,2}\big)$, as $\beta I_H - A_{-1}$ belongs to $\mathcal{L}\big((H,\mathrm{D}(A))_{\theta,2},(H_{-1},H)_{\theta,2}\big)$ by Proposition 1.4.18. We refer to [Lun95, page 134, Theorem 4.3.1(iii)] to obtain that ψ belongs to

$$C^{0,\gamma}\big([0,T];(H,\mathrm{D}(A))_{\theta,2}\big) \cap C^{1,\gamma}\big([0,T];(H_{-1},H)_{\theta,2}\big).$$

It is a result of Proposition 1.4.20 with $\alpha := 1-(\theta-\eta)$ that this space is a subset of

$$C^{[\gamma+\theta-\eta],\gamma+\theta-\eta-[\gamma+\theta-\eta]}\Big([0,T];\big((H_{-1},H)_{\theta,2},(H,\mathrm{D}(A))_{\theta,2}\big)_{1-(\theta-\eta),2}\Big).$$

Now Corollary 1.4.23 and Theorem 1.4.21 imply that

$$\begin{aligned}
&\big((H_{-1},H)_{\theta,2},(H,\mathrm{D}(A))_{\theta,2}\big)_{1-(\theta-\eta),2}\\
&= \big((H_{-1},\mathrm{D}(A))_{\frac{1}{2}\theta,2},(H_{-1},\mathrm{D}(A))_{\frac{1}{2}(\theta+1),2}\big)_{1-(\theta-\eta),2}\\
&= (H_{-1},\mathrm{D}(A))_{\frac{1}{2}(1+\eta)),2} = (H,\mathrm{D}(A))_{\eta,2}.
\end{aligned}$$

Therefore ψ belongs to $C^{[\gamma+\theta-\eta],\gamma+\theta-\eta-[\gamma+\theta-\eta]}\big([0,T];(H,\mathrm{D}(A))_{\eta,2}\big)$. □

Lemma 3.7.4 *Let $f:[0,\infty)\to\mathbb{R}$ belong to $L^p(0,T)$ for some $p\in[1,\infty)$ and every $T>0$. Let ψ be the mild solution to problem (3.59). If there exists $\lambda_0 \geq \beta$ such that $\int_0^\infty \mathrm{e}^{-\lambda_0 t}|f(t)|\,\mathrm{d}t < \infty$, then the Laplace transform $\hat{\psi}:(\lambda_0,\infty)\to(H,\mathrm{D}(A))_{\theta,2}$ is well-defined for every $\theta\in(0,1)$ such that $\theta<\frac{1+\alpha(a)}{2}$, and*

$$\hat{\psi}(\lambda) = \hat{f}(\lambda)(\lambda I_{H_{-1}} - A_{-1})^{-1}(\beta I_H - A_{-1})\pi, \quad \lambda>\lambda_0.$$

Moreover, if a belongs to $\mathcal{K}$ with $\alpha(a)\in(0,1)$, then

$$J\big(\hat{\psi}(\lambda)\big) = \frac{1}{\lambda\hat{a}(\lambda)}\hat{f}(\lambda), \quad \lambda>\lambda_0.$$

PROOF: Let $\theta, \eta \in (0,1)$ be such that $\eta < \theta < \frac{1+\alpha(a)}{2}$. First we show that the integral $\int_0^\infty \mathrm{e}^{-\lambda t}\psi(t)\,\mathrm{d}t$ is Bochner in $(H, \mathrm{D}(A))_{\eta,2}$ for every $\lambda > \lambda_0$. Note that π belongs to $(H, \mathrm{D}(A))_{\theta,2}$ by Proposition 3.5.7, and that by the same arguments as in the proof of Proposition 3.7.2 there exists $M > 0$ such that

$$\left\|(A - \beta I_H)\mathrm{e}^{-\beta t}S(t)\right\|_{\mathcal{L}((H,\mathrm{D}(A))_{\theta,2},(H,\mathrm{D}(A))_{\eta,2})} \le M t^{-1+\theta-\eta}, \quad t > 0.$$

Since the Laplace transform of a convolution is the product of the individual Laplace transforms, and since the mapping $t \mapsto \mathrm{e}^{\beta t} t^{-1+\theta-\eta}$ has Laplace transform $\lambda \mapsto \frac{\Gamma(\theta-\eta)}{(\lambda-\beta)^{\theta-\eta}}$ on (β, ∞), we have

$$\begin{aligned}
&\int_0^\infty \left\|\mathrm{e}^{-\lambda t}\psi(t)\right\|_{(H,\mathrm{D}(A))_{\eta,2}} \mathrm{d}t \\
&\le \int_0^\infty \mathrm{e}^{-\lambda t}\left(\int_0^t \left\|f(s)(\beta I_H - A)S(t-s)\pi\right\|_{(H,\mathrm{D}(A))_{\eta,2}} \mathrm{d}s\right) \mathrm{d}t \\
&\le M\|\pi\|_{(H,\mathrm{D}(A))_{\theta,2}} \int_0^\infty \mathrm{e}^{-\lambda t}\left(\int_0^t \mathrm{e}^{\beta(t-s)}(t-s)^{-1+\theta-\eta}|f(s)|\,\mathrm{d}s\right) \mathrm{d}t \\
&= M\|\pi\|_{(H,\mathrm{D}(A))_{\theta,2}} \frac{\Gamma(\theta-\eta)}{(\lambda-\beta)^{\theta-\eta}} \widehat{|f|}(\lambda) < \infty, \qquad \lambda > \lambda_0.
\end{aligned}$$

Now we use Fubini's theorem and Proposition 1.4.6 to obtain

$$\begin{aligned}
\hat{\psi}(\lambda) &= \int_0^\infty \mathrm{e}^{-\lambda t}\left(\int_0^t f(s)(\beta I_H - A)S(t-s)\pi\,\mathrm{d}s\right) \mathrm{d}t \\
&= \int_0^\infty \mathrm{e}^{-\lambda s} f(s)\left(\int_s^\infty \mathrm{e}^{-\lambda(t-s)}(\beta I_H - A)S(t-s)\pi\,\mathrm{d}t\right) \mathrm{d}s \\
&= \left(\int_0^\infty \mathrm{e}^{-\lambda s} f(s)\,\mathrm{d}s\right)\left(\int_0^\infty \mathrm{e}^{-\lambda t} S_{-1}(t)(\beta I_H - A_{-1})\pi\,\mathrm{d}t\right) \\
&= \hat{f}(\lambda)(\lambda I_{H_{-1}} - A_{-1})^{-1}(\beta I_H - A_{-1})\pi, \qquad \lambda > \lambda_0.
\end{aligned}$$

Finally, we observe that if $\alpha(a) > 0$, then we could have chosen $\eta > \frac{1-\alpha(a)}{2}$, so that J has an extension to $(H, \mathrm{D}(A))_{\eta,2}$ by Proposition 3.5.3. Therefore the last part of the lemma is proved by the following lemma. □

Lemma 3.7.5 *If a belongs to $\mathcal{K}$ with $\alpha(a) \in (0,1)$, then*

$$J\big((\lambda I_{H_{-1}} - A_{-1})^{-1}(\beta I_H - A_{-1})\pi\big) = \frac{1}{\lambda\hat{a}(\lambda)}, \quad \lambda > \beta.$$

PROOF: Let $\lambda > \beta$ and let $\eta \in (0,1)$ be such that $\frac{1-\alpha(a)}{2} < \eta < \frac{1+\alpha(a)}{2}$. Let the sequence $\{\pi_n\}_{n=1}^\infty$ in $\mathrm{D}(A)$ be defined by $\pi_n := n(nI_H - A)^{-1}\pi$ for $n = 1, 2, \cdots$.

Note that the sequence is well-defined by Proposition 3.3.7. Note also that π belongs to $(H, \mathrm{D}(A))_{\eta,2}$ by Proposition 3.5.7. Then it is a consequence of Lemmas 1.4.19 and 1.4.8 that

$$\lim_{n\to\infty} \|\pi_n - \pi\|_{(H,\mathrm{D}(A))_{\eta,2}} = 0.$$

Since $\beta I_{H_{-1}} - A_{-1}$ belongs to $\mathcal{L}\big((H, \mathrm{D}(A))_{\eta,2}, (H_{-1}, H)_{\eta,2}\big)$ and $(\lambda I_{H_{-1}} - A_{-1})^{-1}$ to $\mathcal{L}\big((H_{-1}, H)_{\eta,2}, (H, \mathrm{D}(A))_{\eta,2}\big)$ for every $\lambda \geq \frac{\beta}{4}$ by Proposition 1.4.18 and Lemma 3.3.10, this implies that

$$\lim_{n\to\infty} \|(\beta I_H - A)\pi_n - (\beta I_H - A_{-1})\pi\|_{(H_{-1},H)_{\eta,2}} = 0,$$

and hence,

$$\lim_{n\to\infty} \left\|(\lambda I_H - A)^{-1}(\beta I_H - A)\pi_n - (\lambda I_{H_{-1}} - A_{-1})^{-1}(\beta I_H - A_{-1})\pi\right\|_{(H,\mathrm{D}(A))_{\eta,2}} = 0.$$

As J is continuous on $(H, \mathrm{D}(A))_{\eta,2}$ by Proposition 3.5.3 it follows that

$$\lim_{n\to\infty} J\big((\lambda I_H - A)^{-1}(\beta I_H - A)\pi_n\big) = J\big((\lambda I_{H_{-1}} - A_{-1})^{-1}(\beta I_H - A_{-1})\pi\big). \quad (3.60)$$

To calculate the left-hand side of (3.60) we use Proposition 3.3.7, the definition of π, and Lemmas 3.5.6 and 1.1.5(i) to observe that

$$\begin{aligned}
&J\big((\lambda I_H - A)^{-1}(\beta I_H - A)\pi_n\big) \\
&= \frac{1}{\lambda\hat{a}(\lambda)} \int_{[0,\infty)} \frac{\kappa}{\kappa+\lambda} \big((\beta I_H - A)\pi_n\big)(\kappa)\, \nu(\mathrm{d}\kappa) \\
&= \frac{n}{\lambda\hat{a}(\lambda)} \int_{[0,\infty)} \frac{\kappa}{\kappa+\lambda} \left(\pi(\kappa) + (\beta - n)\frac{\pi(\kappa) + J\big((nI_H - A)^{-1}\pi\big)}{\kappa+n}\right) \nu(\mathrm{d}\kappa) \\
&= \frac{n}{\lambda\hat{a}(\lambda)} \int_{[0,\infty)} \frac{\kappa}{\kappa+\lambda} \left(\frac{(\kappa+\beta)\pi(\kappa) + (\beta - n)J\big((nI_H - A)^{-1}\pi\big)}{\kappa+n}\right) \nu(\mathrm{d}\kappa) \\
&= \frac{1}{\lambda\hat{a}(\lambda)\hat{a}(n)} \int_{[0,\infty)} \frac{\kappa}{(\kappa+\lambda)(\kappa+n)} \nu(\mathrm{d}\kappa) \\
&= \frac{1}{\lambda(n-\lambda)\hat{a}(\lambda)\hat{a}(n)} \int_{[0,\infty)} \left(\frac{n}{\kappa+n} - \frac{\lambda}{\kappa+\lambda}\right) \nu(\mathrm{d}\kappa) \\
&= \frac{n}{\lambda(n-\lambda)\hat{a}(\lambda)} - \frac{1}{(n-\lambda)\hat{a}(n)}, \qquad n = 1, 2, \cdots.
\end{aligned}$$

By Lemmas 1.1.5(i), 1.1.4(i) and (iv), and 1.1.6(i), we have for $n = 1, 2, \cdots$,

$$\begin{aligned}
\hat{a}(n) &= \int_{[0,\infty)} \frac{1}{\kappa+n} \nu(\mathrm{d}\kappa) \leq \int_{[0,1]} \nu(\mathrm{d}\kappa) + \int_{[1,\infty)} \frac{1}{\kappa} \nu(\mathrm{d}\kappa) < \infty, \\
n\hat{a}(n) &= \int_{[0,\infty)} \frac{n}{\kappa+n} \nu(\mathrm{d}\kappa) \leq \int_{[0,\infty)} \nu(\mathrm{d}\kappa) = +\infty.
\end{aligned}$$

Therefore Lebesgue's dominated convergence theorem and the monotone convergence theorem imply that $\lim_{n\to\infty} \hat{a}(n) = 0$ and $\lim_{n\to\infty} n\hat{a}(n) = +\infty$, so that $\lim_{n\to\infty}(n-\lambda)\hat{a}(n) = +\infty$. Hence,

$$\begin{aligned}\lim_{n\to\infty} & J\big((\lambda I_H - A)^{-1}(\beta I_H - A)\pi_n\big) \\ &= \lim_{n\to\infty} \frac{n}{\lambda(n-\lambda)\hat{a}(\lambda)} - \frac{1}{(n-\lambda)\hat{a}(n)} = \frac{1}{\lambda\hat{a}(\lambda)}.\end{aligned}$$

Combined with (3.60) this proves the lemma. □

3.7.2 The scalar linear Volterra equation

In this section we return to the inhomogeneous Volterra equation (3.53), which we are able to solve with the semigroup approach and the results of Section 3.7.1. We conclude with a representation of the resolvent of the first kind of the kernel $a \in \mathcal{K}$.

Theorem 3.7.6 *Let $f : [0,\infty) \to \mathbb{R}$ belong to $L^p(0,T)$ for some $p \in [1,\infty)$ and every $T > 0$. If a belongs to $\mathcal{K}$ with $\alpha(a) \in (0,1)$, then problem (3.53) admits a unique solution u, and u has the representation*

$$u(t) = \begin{cases} \int_0^t f(s) J\big((\beta I_H - A)S(t-s)\pi\big)\,\mathrm{d}s, & t > 0, \\ 0, & t \leq 0. \end{cases} \tag{3.61}$$

Moreover, for all $T > 0$ and $\zeta \in (0, \alpha(a))$ the following holds:

(i) *If $p = 1$, then u belongs to $L^q(0,T)$ for every $q \in \big[1, \frac{1}{1-\zeta}\big)$;*

(ii) *If $p \in \big(1, \frac{1}{\zeta}\big)$, then u belongs to $L^q(0,T)$ where $q := \frac{p}{1-\zeta p}$;*

(iii) *If $p = \frac{1}{\zeta}$, then u belongs to $L^q(0,T)$ for every $q \in [1,\infty)$;*

(iv) *If $p \in \big(\frac{1}{\zeta}, \infty\big)$, then u belongs to $C_0^{0,\zeta-\frac{1}{p}}[0,T]$.*

In particular, if f belongs to $C_0^{0,\gamma}[0,T]$ for some $\gamma \in (0,1)$ and every $T > 0$, then u belongs to

$$C^{[\gamma+\zeta],\gamma+\zeta-[\gamma+\zeta]}[0,T]$$

for all $T > 0$ and $\zeta \in (0,\alpha(a))$ such that $\zeta \neq 1-\gamma$.

PROOF: Let $T > 0$ and $\zeta \in (0,\alpha(a))$, and let $\eta := \frac{1-\zeta}{2}$ and $\theta := \frac{1+\zeta}{2}$. Then we have $\theta - \eta = \zeta$ and $0 < \frac{1-\alpha(a)}{2} < \eta \leq \theta < \frac{1+\alpha(a)}{2} < 1$. Let ψ be the mild solution to problem (3.59). Since $\psi(t)$ has values in $(H, \mathrm{D}(A))_{\eta,2}$ by Proposition 3.7.3,

and J is continuous on $(H, \mathrm{D}(A))_{\eta,2}$ by Proposition 3.5.3, it follows that u given by (3.61) is such that $u(t) = J(\psi(t))$ for every $t > 0$. Even more, the regularity properties of ψ in Proposition 3.7.3 imply those of u.
Now we show that u is the unique solution to problem (3.53). We observe that u satisfies Definition 3.1.1(i), since $u(t) = 0$ for every $t \le 0$ and both a and u belong to $L^1(0,T)$. To show that u satisfies Definition 3.1.1(ii) and (iii) it is sufficient to prove that

$$\int_0^t a(t-s)u(s)\,\mathrm{d}s = \int_0^t f(s)\,\mathrm{d}s, \quad t \in [0,T]. \tag{3.62}$$

To be able to apply the Laplace transform we define the function $f_T : [0,\infty) \to \mathbb{R}$ by

$$f_T(t) := \begin{cases} f(t), & t \in [0,T], \\ 0, & t \in (T,\infty), \end{cases}$$

and let ψ_T denote the mild solution to problem (3.59) with f replaced by f_T. Furthermore, we define $u_T : [0,T] \to \mathbb{R}$ by $u_T(t) := J(\psi_T(t))$ for every $t \in [0,T]$. Note that $\psi_T(t) = \psi(t)$ and $u_T(t) = u(t)$ for every $t \in [0,T]$. Using Lemma 3.7.4 with $\lambda_0 := \beta$ we have

$$\hat{u}_T(\lambda) = \int_0^\infty \mathrm{e}^{-\lambda t} J\big(\psi_T(t)\big)\,\mathrm{d}t = J\big(\hat{\psi}_T(\lambda)\big) = \frac{1}{\lambda \hat{a}(\lambda)} \hat{f}_T(\lambda), \quad \lambda > \beta,$$

that is,

$$\hat{a}(\lambda)\hat{u}_T(\lambda) = \frac{1}{\lambda}\hat{f}_T(\lambda), \quad \lambda > \beta.$$

As the Laplace transform of a convolution is the product of the individual Laplace transforms and the mapping $t \mapsto \int_0^t f(s)\,\mathrm{d}s$ has Laplace transform $\lambda \mapsto \frac{1}{\lambda}\hat{f}(\lambda)$, the inversion theorem for Laplace transforms implies that

$$\int_0^t a(t-s)u_T(s)\,\mathrm{d}s = \int_0^t f_T(s)\,\mathrm{d}s, \quad t > 0.$$

This show that (3.62) holds and the theorem is proved. □

Theorem 3.7.7 *If a belongs to $\mathcal{K}$ with $\alpha(a) \in (0,1)$, then the resolvent b of the first kind of a has the representation*

$$b(t) = J\big((\beta I_H - A)S(t)\pi\big), \quad t > 0. \tag{3.63}$$

If, in addition, $\alpha(a) > \frac{1}{2}$, then b belongs to $L^2(0,T)$ for every $T > 0$.

PROOF: By Theorem 1.3.2 it suffices for the first part of the theorem to show that b given by (3.63) belongs to $L^1(0,T)$ for every $T>0$ and that

$$\int_0^t a(t-s)b(s)\,\mathrm{d}s = 1, \quad t>0. \tag{3.64}$$

Let $\theta,\eta \in (0,1)$ be such that $\frac{1-\alpha(a)}{2} < \eta < \theta < \frac{1+\alpha(a)}{2}$. Then π belongs to $(H,\mathrm{D}(A))_{\theta,2}$ by Proposition 3.5.7, and J is continuous on $(H,\mathrm{D}(A))_{\eta,2}$ by Proposition 3.5.3. Since $A-\beta I_H$ is the infinitesimal generator of the analytic semigroup $\{\mathrm{e}^{-\beta t}S(t)\}_{t\geq 0}$ on H with $0\in\rho(A-\beta I_H)$ by Propositions 1.4.9 and 1.4.13 and Lemma 3.3.10, and since $(H,\mathrm{D}(A))_{\eta,2} = (H,\mathrm{D}(A-\beta I_H))_{\eta,2}$ by Lemma 1.4.15, we have by Corollary 1.4.30(ii) that there exists $M>0$ such that

$$\big\|(A-\beta I_H)\mathrm{e}^{\beta t}S(t)\big\|_{\mathcal{L}((H,\mathrm{D}(A))_{\theta,2},(H,\mathrm{D}(A))_{\eta,2})} \leq M t^{-1+\theta-\eta}, \quad t>0.$$

Therefore we have

$$\begin{aligned}\int_0^T |b(t)|\,\mathrm{d}t &= \int_0^T \big|J\big((\beta I_H - A)S(t)\pi\big)\big|\,\mathrm{d}t \\ &\leq M\mathrm{e}^{\beta T}\|J\|_{\mathcal{L}((H,\mathrm{D}(A))_{\eta,2},\mathbb{R})}\|\pi\|_{(H,\mathrm{D}(A))_{\theta,2}}\int_0^T t^{-1+\theta-\eta}\,\mathrm{d}t < \infty, \qquad T>0.\end{aligned}$$

Furthermore, the Laplace transform $\hat{b}:(\beta,\infty)\to\mathbb{R}$ is well-defined. Indeed, since the mapping $t\mapsto \mathrm{e}^{\beta t}t^{-1+\theta-\eta}$ has Laplace transform $\lambda\mapsto\frac{\Gamma(\theta-\eta)}{(\lambda-\beta)^{\theta-\eta}}$ on (β,∞) we have

$$\begin{aligned}\int_0^\infty \mathrm{e}^{-\lambda t}|b(t)|\,\mathrm{d}t &= \int_0^\infty \mathrm{e}^{-\lambda t}\big|J\big((\beta I_H - A)S(t)\pi\big)\big|\,\mathrm{d}t \\ &\leq M\|J\|_{\mathcal{L}((H,\mathrm{D}(A))_{\eta,2},\mathbb{R})}\|\pi\|_{(H,\mathrm{D}(A))_{\theta,2}}\int_0^\infty \mathrm{e}^{-\lambda t}\mathrm{e}^{\beta t}t^{-1+\theta-\eta}\,\mathrm{d}t \\ &= M\|J\|_{\mathcal{L}((H,\mathrm{D}(A))_{\eta,2},\mathbb{R})}\|\pi\|_{(H,\mathrm{D}(A))_{\theta,2}}\frac{\Gamma(\theta-\eta)}{(\lambda-\beta)^{\theta-\eta}} < \infty, \qquad \lambda>\beta.\end{aligned}$$

Using Proposition 1.4.6 and Lemma 3.7.5 it follows that

$$\begin{aligned}\hat{b}(\lambda) &= \int_0^\infty \mathrm{e}^{-\lambda t}J\big((\beta I_H - A)S(t)\pi\big)\,\mathrm{d}t \\ &= J\left(\int_0^\infty \mathrm{e}^{-\lambda t}S_{-1}(t)(\beta I_H - A_{-1})\pi\,\mathrm{d}t\right) \\ &= J\big((\lambda I_H - A_{-1})^{-1}(\beta I_H - A_{-1})\pi\big) = \frac{1}{\lambda\hat{a}(\lambda)}, \qquad \lambda>\beta,\end{aligned}$$

that is,

$$\hat{a}(\lambda)\hat{u}(\lambda) = \frac{1}{\lambda}, \quad \lambda>\beta.$$

Now the inversion theorem for Laplace transforms implies that (3.64) holds, as the Laplace transform of a convolution is the product of the individual Laplace transforms.

For the second part of the theorem we remark that if $\alpha(a) > \frac{1}{2}$, then we could have chosen θ and η such that $\theta - \eta > \frac{1}{2}$, so that

$$\int_0^T |b(t)|^2 \, \mathrm{d}t = \int_0^T \left|J\big((\beta I_H - A)S(t)\pi\big)\right|^2 \mathrm{d}t$$
$$\leq M^2 \mathrm{e}^{2\beta T} \|J\|^2_{\mathcal{L}((H,\mathrm{D}(A))_{\eta,2},\mathbb{R})} \|\pi\|^2_{(H,\mathrm{D}(A))_{\theta,2}} \int_0^T t^{-2+2(\theta-\eta)} \, \mathrm{d}t < \infty, \quad T > 0.$$

□

Chapter 4

Volterra equations in a separable Hilbert space

In this chapter we study three different types of Hilbert-valued Volterra integrodifferential equations. The first type is the Hilbert-valued equivalent of the scalar linear Volterra equation that was discussed in Chapter 3. The second and third type concern respectively a semilinear Volterra equation and a linear Volterra equation of the first kind.

Throughout the chapter we assume that $(X, \langle\cdot,\cdot\rangle, \|\cdot\|)$ is a real separable Hilbert space.

4.1 Linear Volterra equations

We consider the linear X-valued Volterra integrodifferential equation

$$\begin{aligned} \frac{\mathrm{d}}{\mathrm{d}t}\int_{-\infty}^{t} a(t-s)u(s)\,\mathrm{d}s &= f(t), \quad t>0, \\ u(t) &= u_0(t), \quad t \le 0, \end{aligned} \tag{4.1}$$

where the kernel $a : (0,\infty) \to \mathbb{R}$ belongs to the class $\mathcal{K}$ and where the functions $u_0 : (-\infty, 0] \to X$ and $f : [0,\infty) \to X$ are at least locally integrable. We consider the following notion of solution:

Definition 4.1.1 A solution to problem (4.1) is a strongly Borel measurable function $u : \mathbb{R} \to X$ such that

(i) $\int_{-\infty}^{t} a(t-s)\|u(s)\|\,\mathrm{d}s < \infty$ for every $t \ge 0$;

(ii) The mapping $t \mapsto \int_{-\infty}^{t} a(t-s)u(s)\,\mathrm{d}s$ belongs to $W^{1,1}([0,T];X)$ for every $T > 0$;

(iii) u satisfies (4.1) for almost every $t \in \mathbb{R}$.

We remark that a function $u : \mathbb{R} \to X$ is strongly Borel measurable if and only if the mapping $t \mapsto \langle u(t), x \rangle$ is Borel measurable on $\mathbb{R}$ for every $x \in X$, due to the separability of X, see [DU77, page 42, Theorem 2]. Moreover, analogous to Section 3.1 we have that a solution u to problem (4.1) always belongs to $L^1(0, T; X)$ for every $T > 0$. We also have uniqueness of solutions as shown in the next proposition:

Proposition 4.1.2 *Problem (4.1) admits at most one solution.*

PROOF: Let u be a solution to problem (4.1) with both u_0 and f identically zero. Then Definition 4.1.1(ii) and (iii) imply that $\int_0^t a(t-s)u(s)\,\mathrm{d}s = 0$ for every $t > 0$ and hence,

$$\int_0^t a(t-s)\langle u(s), x \rangle\,\mathrm{d}s = 0, \quad t \geq 0,\, x \in X.$$

It follows from Corollary 1.2.2 that $\langle u(t), x \rangle = 0$ for every $x \in X$ and almost every $t > 0$. Therefore, $u(t) = 0$ for almost every $t > 0$. □

Similar to the scalar equation in Chapter 3 we split (4.1) into a homogeneous equation and an inhomogeneous equation with u_0 identically zero.

4.1.1 The homogeneous problem

We consider the homogeneous Volterra equation

$$\begin{aligned} \frac{\mathrm{d}}{\mathrm{d}t}\int_{-\infty}^t a(t-s)u(s)\,\mathrm{d}s &= 0, \qquad t > 0, \\ u(t) &= u_0(t), \quad t \leq 0, \end{aligned} \tag{4.2}$$

where the kernel $a : (0, \infty) \to \mathbb{R}$ belongs to the class $\mathcal{K}$ and where the function $u_0 : (-\infty, 0] \to X$ satisfies the next hypothesis:

Hypothesis 4.1.3 The function $u_0 : (-\infty, 0] \to X$ is strongly Borel measurable and has the following properties:

(i) There exist $M_1 > 0$ and $\omega > 0$ such that $\|u_0(t)\| \leq M_1 \mathrm{e}^{\omega t}$ for every $t \leq 0$;

(ii) There exist $M_2 > 0$ and $\delta > 0$ such that $\|u_0(0) - u_0(t)\| \leq M_2 |t|$ for every $t \in [-\delta, 0]$.

Let ν denote the unique Borel measure on $[0,\infty)$ such that

$$a(t) = \int_{[0,\infty)} \mathrm{e}^{-\kappa t}\,\nu(\mathrm{d}\kappa), \quad t > 0.$$

Following our semigroup approach we consider problem (4.2) in an abstract setting, that is, the homogeneous abstract Cauchy problem

$$\begin{aligned} \tfrac{\mathrm{d}}{\mathrm{d}t}\psi(t) &= \mathcal{A}\psi(t), \quad t > 0, \\ \psi(0) &= \psi_0, \end{aligned} \tag{4.3}$$

where $\psi_0 : [0,\infty) \to X$ is defined by

$$\psi_0(\kappa) := \int_0^\infty \mathrm{e}^{-\kappa s} u_0(-s)\,\mathrm{d}s, \quad \kappa \geq 0. \tag{4.4}$$

Before we specify this abstract setting we recall some definitions from Chapter 3. The real Hilbert space $(H, \langle\cdot,\cdot\rangle_H, \|\cdot\|_H)$ of equivalence classes is given by

$$H := \Big\{\varphi : [0,\infty) \to \mathbb{R}; \varphi \text{ Borel measurable and } \int_{[0,\infty)} |\varphi(\kappa)|^2(\kappa+\beta)\,\nu(\mathrm{d}\kappa) < \infty\Big\},$$

and endowed with the inner product

$$\langle\varphi,\psi\rangle_H := \int_{[0,\infty)} \varphi(\kappa)\psi(\kappa)(\kappa+\beta)\,\nu(\mathrm{d}\kappa), \quad \varphi,\psi \in H,$$

where $\beta \in \mathbb{R}$ is fixed but arbitrary. The linear functional $J : \mathrm{D}(J) \subseteq H \to \mathbb{R}$ is given by

$$\mathrm{D}(J) := \big\{\varphi \in H; \text{ there exists (a unique) } u \in \mathbb{R} \text{ such that } \kappa \mapsto u - \kappa\varphi(\kappa) \in H\big\},$$

$$J(\varphi) := u, \quad \varphi \in \mathrm{D}(J),$$

and the linear operator $A : \mathrm{D}(A) \subseteq H \to H$ by

$$\mathrm{D}(A) := \left\{\varphi \in \mathrm{D}(J); \int_{[0,\infty)} \big(J(\varphi) - \kappa\varphi(\kappa)\big)\,\nu(\mathrm{d}\kappa) = 0\right\},$$

$$(A\varphi)(\kappa) := J(\varphi) - \kappa\varphi(\kappa), \quad \varphi \in \mathrm{D}(A),\ \kappa \geq 0.$$

According to Theorems 3.3.8 and 3.3.11 we have that A is the infinitesimal generator of an analytic semigroup $\{S(t)\}_{t\geq 0}$ on H. Moreover, the linear operator $A - \frac{\beta}{4} I_H$ is m-dissipative in $(H, \langle\cdot,\cdot\rangle_H)$ by Proposition 3.3.7.

We shall consider the homogeneous abstract Cauchy problem (4.3) in the real Hilbert space $(\mathcal{H}, \langle\cdot,\cdot\rangle_{\mathcal{H}})$ of equivalence classes defined by

$$\mathcal{H} := \Big\{ g : [0,\infty) \to X; g \text{ strongly Borel measurable and} \\ \int_{[0,\infty)} \|g(\kappa)\|^2 (\kappa+\beta)\,\nu(\mathrm{d}\kappa) < \infty \Big\},$$

endowed with the inner product

$$\langle g, h\rangle_{\mathcal{H}} := \int_{[0,\infty)} \big\langle g(\kappa), h(\kappa)\big\rangle (\kappa+\beta)\,\nu(\mathrm{d}\kappa), \quad g,h \in \mathcal{H}.$$

For $\varphi \in H$ and $x \in X$ we define the function $\varphi \otimes x \in \mathcal{H}$ by

$$(\varphi \otimes x)(\kappa) := \varphi(\kappa)x, \quad \kappa \geq 0.$$

The real vector space of all linear combinations of $\varphi \otimes x$ is denoted by $H \otimes X$, that is,

$$H \otimes X := \left\{ \sum_{i=1}^{n} (\varphi_i \otimes x_i);\ n \in \mathbb{N},\ \varphi_1, \cdots, \varphi_n \in H,\ x_1, \cdots, x_n \in X \right\}.$$

Note that $H \otimes X$ is a dense subset of $\mathcal{H}$ since $H \otimes X$ contains the simple functions of $\mathcal{H}$, and that

$$\|\varphi \otimes x\|_{\mathcal{H}} = \|\varphi\|_H \|x\|, \quad \varphi \in H,\ x \in X.$$

Theorem 4.1.4 *There exists a unique analytic semigroup $\{\mathcal{S}(t)\}_{t\geq 0}$ on $\mathcal{H}$ such that for every $t \geq 0$,*

$$\mathcal{S}(t)(\varphi \otimes x) = S(t)\varphi \otimes x, \quad \varphi \in H,\ x \in X,$$

and

$$\|\mathcal{S}(t)\|_{\mathcal{L}(\mathcal{H})} = \|S(t)\|_{\mathcal{L}(H)}.$$

Furthermore, the infinitesimal generator $\mathcal{A} : \mathrm{D}(\mathcal{A}) \subseteq \mathcal{H} \to \mathcal{H}$ of $\{\mathcal{S}(t)\}_{t\geq 0}$ is such that $\mathcal{A} - \frac{\beta}{4} I_{\mathcal{H}}$ is m-dissipative in $(\mathcal{H}, \langle\cdot,\cdot\rangle_{\mathcal{H}})$ and for every $\lambda > \frac{\beta}{4}$,

$$(\lambda I_{\mathcal{H}} - \mathcal{A})^{-1}(\varphi \otimes x) = (\lambda I_H - A)^{-1}\varphi \otimes x, \quad \varphi \in H,\ x \in X,$$

and

$$\|(\lambda I_{\mathcal{H}} - \mathcal{A})^{-1}\|_{\mathcal{L}(\mathcal{H})} = \|(\lambda I_H - A)^{-1}\|_{\mathcal{L}(H)}.$$

The extension in Theorem 4.1.4 is based on the following Marcinkiewicz-Zygmund principle, see for example [EG77, page 203, Theorem] and [Egb92, page 32].

Theorem 4.1.5 *Let $(\Omega, \mathcal{F}, \mu)$ be a σ-finite measure space and E a real Hilbert space. Let $T : L^p(\Omega) \to L^p(\Omega)$ be a bounded linear operator for some $p \in [1, \infty)$. Then T is uniquely extensible to a bounded linear operator $\mathcal{T} : L^p(\Omega; E) \to L^p(\Omega; E)$ such that*

$$\mathcal{T}(\varphi \otimes x) = T\varphi \otimes x, \quad \varphi \in L^p(\Omega),\, x \in E.$$

Moreover,

$$\|\mathcal{T}\|_{\mathcal{L}(L^p(\Omega;E))} = \|T\|_{\mathcal{L}(L^p(\Omega))}.$$

In order to find explicit formulations for the domain and action of $\mathcal{A}$ we define the linear functional $\mathcal{J} : \mathrm{D}(\mathcal{J}) \subseteq \mathcal{H} \to X$ by

$$\mathrm{D}(\mathcal{J}) := \big\{ g \in \mathcal{H}; \text{there exists (a unique) } y \in X \text{ such that } \kappa \mapsto y - \kappa g(\kappa) \in \mathcal{H} \big\},$$

$$\mathcal{J}(g) := y, \quad g \in \mathrm{D}(\mathcal{J}).$$

Well-definedness of $\mathcal{J}$ follows from the next lemma, the proof of which is analogous to the one of Lemma 3.3.3. Note that $\mathrm{D}(J) \otimes X \subseteq \mathrm{D}(\mathcal{J})$ and $\mathcal{J}(\varphi \otimes x) = J(\varphi)x$ for all $\varphi \in \mathrm{D}(J)$ and $x \in X$.

Lemma 4.1.6 *For every $g \in \mathcal{H}$ there exists at most one $y \in X$ such that the mapping $\kappa \mapsto y - \kappa g(\kappa)$ belongs to $\mathcal{H}$.*

Let the linear operator $\tilde{\mathcal{A}} : \mathrm{D}(\tilde{\mathcal{A}}) \subseteq \mathcal{H} \to \mathcal{H}$ be defined by

$$\mathrm{D}(\tilde{\mathcal{A}}) := \left\{ g \in \mathrm{D}(\mathcal{J});\ \int_{[0,\infty)} \big(\mathcal{J}(g) - \kappa g(\kappa)\big)\, \nu(\mathrm{d}\kappa) = 0 \right\},$$

$$(\tilde{\mathcal{A}} g)(\kappa) := \mathcal{J}(g) - \kappa g(\kappa), \quad g \in \mathrm{D}(\tilde{\mathcal{A}}),\, \kappa \geq 0.$$

Well-definedness of $\tilde{\mathcal{A}}$ follows from the lemma below that is proved analogously to Lemma 3.3.5. Note that $\mathrm{D}(A) \otimes X \subseteq \mathrm{D}(\tilde{\mathcal{A}})$ and $\tilde{\mathcal{A}}(\varphi \otimes x) = A\varphi \otimes x$ for all $\varphi \in \mathrm{D}(A)$ and $x \in X$.

Lemma 4.1.7 *The Hilbert space $\mathcal{H}$ is continuously embedded in the real Banach space $L^1([0,\infty), \nu; X)$ with*

$$\|g\|_{L^1([0,\infty),\nu;X)} \leq \sqrt{\hat{a}(\beta)}\, \|g\|_{\mathcal{H}}, \quad g \in \mathcal{H}.$$

We shall prove that $\tilde{\mathcal{A}} = \mathcal{A}$. For this purpose we state the following two propositions; the proof of the latter is analogous to the one of Proposition 3.3.7.

Proposition 4.1.8 *The linear operator $\tilde{\mathcal{A}} - \frac{\beta}{4} I_{\mathcal{H}} : \mathrm{D}(\tilde{\mathcal{A}}) \subseteq \mathcal{H} \to \mathcal{H}$ is dissipative in $(\mathcal{H}, \langle \cdot, \cdot \rangle_{\mathcal{H}})$.*

PROOF: Using the definition of $\mathrm{D}(\tilde{\mathcal{A}})$ and Lemmas 4.1.7 and 1.1.5(i), we have

$$
\begin{aligned}
\langle \tilde{\mathcal{A}} g, g\rangle_{\mathcal{H}} &= \int_{[0,\infty)} \langle \mathcal{J}(g) - \kappa g(\kappa), g(\kappa)\rangle(\kappa+\beta)\,\nu(\mathrm{d}\kappa) - \\
&\quad \int_{[0,\infty)} \langle \mathcal{J}(g) - \kappa g(\kappa), \mathcal{J}(g)\rangle\,\nu(\mathrm{d}\kappa) \\
&= \int_{[0,\infty)} \left\langle \frac{\kappa}{\kappa+\beta}\mathcal{J}(g) - \kappa g(\kappa), (\kappa+\beta)g(\kappa) - \mathcal{J}(g)\right\rangle \nu(\mathrm{d}\kappa) + \\
&\quad \int_{[0,\infty)} \left\langle \frac{\beta}{\kappa+\beta}\mathcal{J}(g), (\kappa+\beta)g(\kappa)\right\rangle \nu(\mathrm{d}\kappa) - \\
&\quad \int_{[0,\infty)} \left\langle \frac{\beta}{\kappa+\beta}\mathcal{J}(g), \mathcal{J}(g)\right\rangle \nu(\mathrm{d}\kappa) \\
&= -\int_{[0,\infty)} \frac{\kappa}{\kappa+\beta}\left\|\mathcal{J}(g) - (\kappa+\beta)g(\kappa)\right\|^2 \nu(\mathrm{d}\kappa) + \\
&\quad \beta\left\langle \mathcal{J}(g), \int_{[0,\infty)} g(\kappa)\,\nu(\mathrm{d}\kappa)\right\rangle - \beta\hat{a}(\beta)\|\mathcal{J}(g)\|^2 \\
&\le \beta\sqrt{\hat{a}(\beta)}\,\|\mathcal{J}(g)\|\,\|g\|_{\mathcal{H}} - \beta\hat{a}(\beta)\|\mathcal{J}(g)\|^2 \\
&= -\beta\big(\sqrt{\hat{a}(\beta)}\,\|\mathcal{J}(g)\| - \tfrac{1}{2}\|g\|_{\mathcal{H}}\big)^2 + \tfrac{\beta}{4}\|g\|_{\mathcal{H}}^2 \le \tfrac{\beta}{4}\|g\|_{\mathcal{H}}^2, \quad g \in \mathrm{D}(\tilde{\mathcal{A}}).
\end{aligned}
$$

This implies that $\big\langle(\tilde{\mathcal{A}} - \frac{\beta}{4}I_{\mathcal{H}})g, g\big\rangle_{\mathcal{H}} \le 0$ for every $g \in \mathrm{D}(\tilde{\mathcal{A}})$. □

Proposition 4.1.9 *The linear operator $\tilde{\mathcal{A}} - \frac{\beta}{4}I_{\mathcal{H}} : \mathrm{D}(\tilde{\mathcal{A}}) \subseteq \mathcal{H} \to \mathcal{H}$ is m-dissipative in $(\mathcal{H}, \langle\cdot,\cdot\rangle_{\mathcal{H}})$. Moreover, if $\lambda > \frac{\beta}{4}$ and g belongs to $\mathcal{H}$, then*

$$
\big((\lambda I_{\mathcal{H}} - \tilde{\mathcal{A}})^{-1} g\big)(\kappa) = \frac{1}{\kappa+\lambda}(g(\kappa) + y), \quad \kappa \ge 0,
$$

where $y \in X$ is defined by

$$
y := \frac{1}{\lambda\hat{a}(\lambda)} \int_{[0,\infty)} \frac{\kappa}{\kappa+\lambda} g(\kappa)\,\nu(\mathrm{d}\kappa).
$$

In particular,

$$
y = \mathcal{J}\big((\lambda I_{\mathcal{H}} - \tilde{\mathcal{A}})^{-1} g\big),
$$

and

$$
(\lambda I_{\mathcal{H}} - \tilde{\mathcal{A}})^{-1}(\varphi \otimes x) = (\lambda I_H - A)^{-1}\varphi \otimes x, \quad \varphi \in H,\ x \in X.
$$

Proposition 4.1.10 $\mathcal{A} = \tilde{\mathcal{A}}$.

PROOF: Let $\lambda > \frac{\beta}{4}$. Theorem 4.1.4 and Proposition 4.1.9 imply that the bounded linear operators $(\lambda I_{\mathcal{H}} - \mathcal{A})^{-1}$ and $(\lambda I_{\mathcal{H}} - \tilde{\mathcal{A}})^{-1}$ coincide on $H \otimes X$. Since $H \otimes X$ is dense in $\mathcal{H}$ it follows that $(\lambda I_{\mathcal{H}} - \mathcal{A})^{-1} = (\lambda I_{\mathcal{H}} - \tilde{\mathcal{A}})^{-1}$ and hence, $\mathcal{A} = \tilde{\mathcal{A}}$. □

At this point we have specified the abstract setting for the Volterra equation (4.2), that is, the homogeneous abstract Cauchy problem (4.3) in $\mathcal{H}$. The next lemma states that ψ_0 belongs to $\mathcal{H}$, and more. The proof of this lemma is analogous to the proof of Lemma 3.4.4 combined with Lemma 3.4.2.

Lemma 4.1.11 *If u_0 satisfies Hypothesis 4.1.3, then ψ_0 belongs to $\mathrm{D}(\mathcal{J})$ and $\mathcal{J}(\psi_0) = u_0(0)$. In particular, ψ_0 belongs to $\mathrm{D}(\mathcal{A})$ if and only if u_0 also satisfies*

$$\frac{\mathrm{d}^-}{\mathrm{d}t}\bigg|_{t=0} \int_{-\infty}^{t} a(t-s)u_0(s)\,\mathrm{d}s = 0, \tag{4.5}$$

or equivalently,

$$a(+\infty)u_0(0) + \int_0^\infty \big(-\tfrac{\mathrm{d}}{\mathrm{d}s}a(s)\big)\big(u_0(0) - u_0(-s)\big)\,\mathrm{d}s = 0.$$

Now we are in a position to solve problem (4.3) in the following sense:

Definition 4.1.12 A *strict solution* to problem (4.3) is a function $\psi : [0,\infty) \to \mathcal{H}$ such that ψ belongs to $C\big([0,\infty);\mathrm{D}(\mathcal{A})\big) \cap C^1\big([0,\infty);\mathcal{H}\big)$, where $\mathrm{D}(\mathcal{A})$ is endowed with the graph norm of $\mathcal{A}$, and such that ψ satisfies (4.3) for every $t \geq 0$.

Proposition 4.1.13 *If u_0 satisfies Hypothesis 4.1.3 and (4.5), then problem (4.3) admits a unique strict solution ψ. Moreover, ψ has the representation $\psi(t) = \mathcal{S}(t)\psi_0$ for every $t \geq 0$ and ψ is real analytic in $(0,\infty)$ with respect to the graph norm of $\mathcal{A}$. In particular, ψ belongs to $C^{0,1-\eta}\big([0,T];(\mathcal{H},\mathrm{D}(\mathcal{A}))_{\eta,2}\big)$ for all $T > 0$ and $\eta \in (0,1)$.*

PROOF: We refer to [Lun95, page 126, Lemma 4.1.6] and use the analyticity of the semigroup combined with Propositions 1.4.29(ii)and 1.4.31(ii). □

With Proposition 4.1.13 we are able to solve problem (4.2). The theorem below is proved analogously to Theorem 3.4.8. However, we need the following lemma, proved in the same way as Lemma 3.3.12. We denote the graph norm of $\mathcal{A}$ by $\|\cdot\|_{\mathcal{A}}$.

Lemma 4.1.14 *The restriction $\mathcal{J}|_{\mathrm{D}(\mathcal{A})} : (\mathrm{D}(\mathcal{A}), \|\cdot\|_{\mathcal{A}}) \to X$ is continuous.*

Theorem 4.1.15 *Let u_0 satisfy Hypothesis 4.1.3 and (4.5). Then problem (4.2) admits a unique solution u. Moreover, u is continuous in $[0,\infty)$, real analytic in $(0,\infty)$, and has the representation*

$$u(t) = \begin{cases} \mathcal{J}\big(\mathcal{S}(t)\psi_0\big), & t > 0, \\ u_0(t), & t \le 0, \end{cases} \tag{4.6}$$

where ψ_0 is given by (4.4).

Proposition 4.1.13 implies that if $\mathcal{J}|_{\mathrm{D}(\mathcal{A})}$ is extensible to a bounded linear operator from $(\mathcal{H}, \mathrm{D}(\mathcal{A}))_{\eta,2}$ into X for some $\eta \in (0,1)$, then the solution u in Theorem 4.1.15 has more regularity, namely u belongs to $C^{0,1-\eta}([0,T];X)$ for every $T > 0$. To obtain this extension of $\mathcal{J}|_{\mathrm{D}(\mathcal{A})}$ we need explicit formulations for the domain and action of the adjoint operator $\mathcal{A}^*$ of $\mathcal{A}$ and a characterization of its resolvent operator. This is stated in the lemma and proposition below. The latter is proved analogously to Propositions 4.1.9 and 3.3.14 using Lemma 1.1.5(vi).

Lemma 4.1.16 *The adjoint operator $\mathcal{A}^* : \mathrm{D}(\mathcal{A}^*) \subseteq \mathcal{H} \to \mathcal{H}$ is given by*

$$\mathrm{D}(\mathcal{A}^*) = \left\{ g \in \mathrm{D}(\mathcal{J});\ \int_{[0,\infty)} \big(\mathcal{J}(g) - (\kappa+\beta) g(\kappa)\big)\, \nu(\mathrm{d}\kappa) = 0 \right\},$$

$$(\mathcal{A}^* g)(\kappa) = \frac{\kappa}{\kappa+\beta}\mathcal{J}(g) - \kappa g(\kappa), \quad g \in \mathrm{D}(\mathcal{A}^*),\ \kappa \ge 0.$$

PROOF: Let g belong to $\mathcal{H}$. Then g belongs to $\mathrm{D}(\mathcal{A}^*) = \mathrm{D}(\beta I_{\mathcal{H}} - \mathcal{A}^*)$ if there exists $h \in \mathcal{H}$ such that

$$\langle k, g\rangle_{\mathcal{H}} = \big\langle (\beta I_{\mathcal{H}} - \mathcal{A})^{-1} k, h \big\rangle_{\mathcal{H}}, \quad k \in \mathcal{H}. \tag{4.7}$$

We fix an arbitrary $k \in \mathcal{H}$ and let $y \in X$ be given by

$$y := \frac{1}{\beta \hat{a}(\beta)} \int_{[0,\infty)} \frac{\kappa}{\kappa+\beta} k(\kappa)\, \nu(\mathrm{d}\kappa).$$

Using Proposition 4.1.9 we have that (4.7) holds if and only if

$$\int_{[0,\infty)} \langle k(\kappa), g(\kappa)\rangle (\kappa+\beta)\, \nu(\mathrm{d}\kappa) = \int_{[0,\infty)} \langle k(\kappa) + y, h(\kappa)\rangle\, \nu(\mathrm{d}\kappa),$$

if and only if

$$\begin{aligned} &\int_{[0,\infty)} \left\langle k(\kappa), g(\kappa) - \frac{1}{\kappa+\beta} h(\kappa) \right\rangle (\kappa+\beta)\, \nu(\mathrm{d}\kappa) \\ &= \left\langle \frac{1}{\beta\hat{a}(\beta)} \int_{[0,\infty)} \frac{\kappa}{\kappa+\beta} k(\kappa)\, \nu(\mathrm{d}\kappa), \int_{[0,\infty)} h(\kappa)\, \nu(\mathrm{d}\kappa) \right\rangle \\ &= \int_{[0,\infty)} \left\langle k(\kappa), \frac{1}{\beta\hat{a}(\beta)} \frac{\kappa}{(\kappa+\beta)^2} \int_{[0,\infty)} h(\tilde{\kappa})\, \nu(\mathrm{d}\tilde{\kappa}) \right\rangle (\kappa+\beta)\, \nu(\mathrm{d}\kappa). \end{aligned} \tag{4.8}$$

Recognizing the inner product of $\mathcal{H}$ in the first and last line in (4.8), it follows from the fact that $k \in \mathcal{H}$ is arbitrary that (4.7) holds if and only if

$$g(\kappa) - \frac{1}{\kappa+\beta} h(\kappa) = \frac{1}{\beta \hat{a}(\beta)} \frac{\kappa}{(\kappa+\beta)^2} \int_{[0,\infty)} h(\tilde{\kappa})\, \nu(\mathrm{d}\tilde{\kappa}), \quad \kappa \geq 0. \tag{4.9}$$

Similar to the proof of Lemma 3.3.13 we have that (4.9) holds if and only if there exists $K \in X$ satisfying

$$\int_{[0,\infty)} \big(K - (\kappa+\beta) g(\kappa)\big)\, \nu(\mathrm{d}\kappa) = 0 \tag{4.10}$$

such that

$$h(\kappa) = (\kappa+\beta) g(\kappa) - \frac{\kappa}{\kappa+\beta} K, \quad \kappa \geq 0. \tag{4.11}$$

Since g and the mapping $\kappa \mapsto \frac{1}{\kappa+\beta}$ belong to $\mathcal{H}$, it follows from (4.11) that h belongs to $\mathcal{H}$ if and only if the mapping $\kappa \mapsto \kappa g(\kappa) - K$ belongs to $\mathcal{H}$. We conclude that g belongs to $\mathrm{D}(\mathcal{A}^*)$ if and only if g belongs to $\mathrm{D}(\mathcal{J})$ and if g satisfies (4.10) with K replaced by $\mathcal{J}(g)$. Now the rest of the lemma follows from straightforward calculation. □

Proposition 4.1.17 *The linear operator $\mathcal{A}^* - \frac{\beta}{4} I_{\mathcal{H}} : \mathrm{D}(\mathcal{A}^*) \subseteq \mathcal{H} \to \mathcal{H}$ is m-dissipative in $(\mathcal{H}, \langle\cdot,\cdot\rangle_{\mathcal{H}})$. Moreover, if $\lambda > \frac{\beta}{4}$ and h belongs to $\mathcal{H}$, then*

$$\big((\lambda I_{\mathcal{H}} - \mathcal{A}^*)^{-1} h\big)(\kappa) = \frac{1}{\kappa+\lambda} \left(h(\kappa) + \frac{\kappa}{\kappa+\beta} K \right), \quad \kappa \geq 0,$$

where $K \in X$ is defined by

$$K := \frac{1}{\lambda \hat{a}(\lambda)} \int_{[0,\infty)} \frac{\kappa+\beta}{\kappa+\lambda} h(\kappa)\, \nu(\mathrm{d}\kappa).$$

In particular,

$$K = \mathcal{J}\big((\lambda I_{\mathcal{H}} - \mathcal{A}^*)^{-1} h\big),$$

and

$$(\lambda I_{\mathcal{H}} - \mathcal{A}^*)^{-1} (\varphi \otimes x) = (\lambda I_H - A^*)^{-1} \varphi \otimes x, \quad \varphi \in H,\, x \in X.$$

Note that $\mathrm{D}(A^*) \otimes X \subseteq \mathrm{D}(\mathcal{A}^*)$ and $\mathcal{A}^*(\varphi \otimes x) = A^* \varphi \otimes x$ for all $\varphi \in \mathrm{D}(A^*)$ and $x \in X$. We recall that the function $\xi \in H$ in Chapter 3 is given by

$$\xi(\kappa) := \frac{\kappa}{(\kappa+\beta)^2}, \quad \kappa \geq 0.$$

Since ξ belongs to $\big(H, \mathrm{D}(A^* - \beta I_H)\big)_{1-\eta,2}$ for every $\eta \in (0,1)$ such that $\eta > \frac{1-\alpha(a)}{2}$ by Proposition 3.5.2, it follows that $\xi \otimes x$ belongs to $\big(\mathcal{H}, \mathrm{D}(\mathcal{A}^* - \beta I_{\mathcal{H}})\big)_{1-\eta,2}$ for every $x \in X$, and

$$\|\xi \otimes x\|_{(\mathcal{H}, \mathrm{D}(\mathcal{A}^* - \beta I_{\mathcal{H}}))_{1-\eta,2}} = \|\xi\|_{(H, \mathrm{D}(A^* - \beta I_H))_{1-\eta,2}} \|x\|, \quad x \in X. \tag{4.12}$$

Proposition 4.1.18 *The restriction $\mathcal{J}|_{\mathrm{D}(\mathcal{A})}$ is uniquely extensible to a bounded linear operator from $(\mathcal{H},\mathrm{D}(\mathcal{A}))_{\eta,2}$ into X for every $\eta\in(0,1)$ such that $\eta>\frac{1-\alpha(a)}{2}$.*

PROOF: Let $\eta\in(0,1)$ be such that $\eta>\frac{1-\alpha(a)}{2}$. We observe that $\mathrm{D}(\mathcal{A})$ is dense in $(\mathcal{H},\mathrm{D}(\mathcal{A}))_{\eta,2}$ by Lemma 1.4.17 and that

$$(\mathcal{H},\mathrm{D}(\mathcal{A}))_{\eta,2}=\big(\mathcal{H},\mathrm{D}(\mathcal{A}-\beta I_{\mathcal{H}})\big)_{\eta,2}=\mathrm{D}\left((\beta I_{\mathcal{H}}-\mathcal{A})^{\eta}\right)$$

by Lemma 1.4.15 and Theorem 1.4.27. Moreover, the norms of $(\mathcal{H},\mathrm{D}(\mathcal{A}))_{\eta,2}$, $\big(\mathcal{H},\mathrm{D}(\mathcal{A}-\beta I_{\mathcal{H}})\big)_{\eta,2}$, and the norm $g\mapsto\|(\beta I_{\mathcal{H}}-\mathcal{A})^{\eta}g\|_{\mathcal{H}}$ on $\mathrm{D}((\beta I_{\mathcal{H}}-\mathcal{A})^{\eta})$ are equivalent. Therefore it is sufficient to show that there exists $M>0$ such that

$$\|\mathcal{J}(g)\|\le M\left\|(\beta I_{\mathcal{H}}-\mathcal{A})^{\eta}g\right\|_{\mathcal{H}},\quad g\in\mathrm{D}(\mathcal{A}). \tag{4.13}$$

We shall use that

$$\|\mathcal{J}(g)\|=\sup_{\substack{x\in X\\ \|x\|=1}}\left|\langle\mathcal{J}(g),x\rangle\right|,\quad g\in\mathrm{D}(\mathcal{A}).$$

Let $x\in X$. Since $\big(\mathcal{H},\mathrm{D}(\mathcal{A}^*-\beta I_{\mathcal{H}})\big)_{1-\eta,2}=\mathrm{D}\left((\beta I_{\mathcal{H}}-\mathcal{A}^*)^{1-\eta}\right)$ with the norm of $\big(\mathcal{H},\mathrm{D}(\mathcal{A}^*-\beta I_{\mathcal{H}})\big)_{1-\eta,2}$ equivalent to the norm $g\mapsto\left\|(\beta I_{\mathcal{H}}-\mathcal{A}^*)^{1-\eta}g\right\|_{\mathcal{H}}$ on $\mathrm{D}\left((\beta I_{\mathcal{H}}-\mathcal{A}^*)^{1-\eta}\right)$, there exists $C>0$ such that, using (4.12),

$$\left\|(\beta I_{\mathcal{H}}-\mathcal{A}^*)^{1-\eta}(\xi\otimes x)\right\|_{\mathcal{H}}\le C\,\|\xi\|_{(H,\mathrm{D}(A^*-\beta I_H))_{1-\eta,2}}\,\|x\|. \tag{4.14}$$

Using Propositions 4.1.9 and 1.4.26 and Lemma 1.4.28, we have

$$\begin{aligned}
\langle\mathcal{J}(g),x\rangle&=\big\langle\big(\mathcal{J}\circ(\beta I_{\mathcal{H}}-\mathcal{A})^{-1}\circ(\beta I_{\mathcal{H}}-\mathcal{A})\big)(g),x\big\rangle\\
&=\left\langle\frac{1}{\beta\hat{a}(\beta)}\int_{[0,\infty)}\frac{\kappa}{\kappa+\beta}\big((\beta I_{\mathcal{H}}-\mathcal{A})g\big)(\kappa)\,\nu(\mathrm{d}\kappa),x\right\rangle\\
&=\frac{1}{\beta\hat{a}(\beta)}\int_{[0,\infty)}\big\langle\big((\beta I_{\mathcal{H}}-\mathcal{A})g\big)(\kappa),\xi(\kappa)x\big\rangle(\kappa+\beta)\,\nu(\mathrm{d}\kappa)\\
&=\frac{1}{\beta\hat{a}(\beta)}\big\langle(\beta I_{\mathcal{H}}-\mathcal{A})g,\xi\otimes x\big\rangle_{\mathcal{H}}\\
&=\frac{1}{\beta\hat{a}(\beta)}\big\langle(\beta I_{\mathcal{H}}-\mathcal{A})g,(\beta I_{\mathcal{H}}-\mathcal{A}^*)^{\eta-1}(\beta I_{\mathcal{H}}-\mathcal{A}^*)^{1-\eta}(\xi\otimes x)\big\rangle_{\mathcal{H}}\\
&=\frac{1}{\beta\hat{a}(\beta)}\big\langle(\beta I_{\mathcal{H}}-\mathcal{A})^{\eta}g,(\beta I_{\mathcal{H}}-\mathcal{A}^*)^{1-\eta}(\xi\otimes x)\big\rangle_{\mathcal{H}},\quad g\in\mathrm{D}(\mathcal{A}).
\end{aligned}$$

Hence, using (4.14) we obtain

$$\begin{aligned}
\left|\langle\mathcal{J}(g),x\rangle\right|&\le\frac{1}{\beta\hat{a}(\beta)}\left\|(\beta I_{\mathcal{H}}-\mathcal{A})^{\eta}g\right\|_{\mathcal{H}}\left\|(\beta I_{\mathcal{H}}-\mathcal{A}^*)^{1-\eta}(\xi\otimes x)\right\|_{\mathcal{H}}\\
&\le\frac{C}{\beta\hat{a}(\beta)}\left\|(\beta I_{\mathcal{H}}-\mathcal{A})^{\eta}g\right\|_{\mathcal{H}}\|\xi\|_{(H,\mathrm{D}(A^*-\beta I_H))_{1-\eta,2}}\,\|x\|,\quad g\in\mathrm{D}(\mathcal{A}).
\end{aligned}$$

This shows that (4.13) holds with $M := \frac{C}{\beta\hat{a}(\beta)}\|\xi\|_{(H,\mathrm{D}(A^*-\beta I_H))_{1-\eta,2}}$. □

Finally we are in a position to state the corollary to Theorem 4.1.15 concerning more regularity of the solution u to problem (4.2). Even more, we do not need the extra condition (4.5) on u_0 for existence of a solution to problem (4.2) as long as $\alpha(a) \in (0,1)$. This is stated in the next theorem. The succeeding corollary is proved in the same way as Corollary 3.6.5.
We recall that the function $\pi \in H$ in Chapter 3 is given by

$$\pi(\kappa) := \frac{1}{\beta\hat{a}(\beta)}\frac{1}{\kappa+\beta}, \quad \kappa \geq 0, \tag{4.15}$$

and that π belongs to $(H,\mathrm{D}(A))_{\theta,2}$ for every $\theta \in (0,1)$ such that $\theta < \frac{1+\alpha(a)}{2}$ by Proposition 3.5.7. Note that $(H,\mathrm{D}(A))_{\theta,2} \otimes X \subseteq (\mathcal{H},\mathrm{D}(\mathcal{A}))_{\theta,2}$ for every $\theta \in (0,1)$ and that

$$\|\varphi \otimes x\|_{(\mathcal{H},\mathrm{D}(\mathcal{A}))_{\theta,2}} = \|\varphi\|_{(H,\mathrm{D}(A))_{\theta,2}}\|x\|, \quad \varphi \in (H,\mathrm{D}(A))_{\theta,2},\ x \in X.$$

Corollary 4.1.19 *If u_0 satisfies Hypothesis 4.1.3 and (4.5), then the unique solution u to problem (4.2) belongs to $C^{0,1-\eta}([0,T];X)$ for all $T > 0$ and $\eta \in (0,1)$ such that $\eta > \frac{1-\alpha(a)}{2}$.*

Theorem 4.1.20 *Let u_0 satisfy Hypothesis 4.1.3. If a belongs to $\mathcal{K}$ with $\alpha(a) \in (0,1)$, then problem (4.2) admits a unique solution u. Moreover, u is real analytic in $(0,\infty)$, has the representation (4.6), and belongs to $C^{0,\zeta}([0,T];X)$ for all $T > 0$ and $\zeta \in [0,\alpha(a))$.*

PROOF: Let $T > 0$ and $\zeta \in [0,\alpha(a))$, and let $\eta := \frac{1-\zeta}{2}$ and $\theta := \frac{1+\zeta}{2}$. Then we have $\theta - \eta = \zeta$ and $0 < \frac{1-\alpha(a)}{2} < \eta \leq \theta < \frac{1+\alpha(a)}{2} < 1$. Let the functions $u_1, u_2 : (-\infty,0] \to X$ and $\psi_1, \psi_2 : [0,\infty) \to X$ be defined by respectively $u_1(t) := u_0(t) - \mathrm{e}^{\beta t}x$ and $u_2(t) := \mathrm{e}^{\beta t}x$ for every $t \leq 0$, and

$$\psi_j(\kappa) := \int_0^\infty \mathrm{e}^{-\kappa s}u_j(-s)\,\mathrm{d}s, \quad \kappa \geq 0,\ j = 1,2,$$

where

$$x := \frac{1}{\beta\hat{a}(\beta)}\left.\frac{\mathrm{d}^-}{\mathrm{d}t}\right|_{t=0}\int_{-\infty}^t a(t-s)u_0(s)\,\mathrm{d}s.$$

Note that u_1 and u_2 satisfy Hypothesis 4.1.3. Even more, u_1 also satisfies (4.5), which is proved just as in the proof of Theorem 3.6.4 using Lemma 3.4.2. At the one hand, we obtain from Theorem 4.1.15 that the function $\tilde{u}_1 : \mathbb{R} \to X$ given by

$$\tilde{u}_1(t) := \begin{cases} \mathcal{J}(\mathcal{S}(t)\psi_1), & t > 0, \\ u_1(t), & t \leq 0, \end{cases}$$

is the unique solution to problem (4.2) with u_0 replaced by u_1, and $\tilde{u}_1$ is real analytic in $(0,\infty)$. Moreover, it follows from Corollary 4.1.19 that $\tilde{u}_1$ belongs to $C^{0,1-\eta}([0,T];X)$. At the other hand, using the definition of π it is a result of Theorem 3.6.4 that the scalar version of problem (4.2) with u_0 replaced by the mapping $t \mapsto \mathrm{e}^{\beta t}$ admits the unique solution

$$t \mapsto \begin{cases} \beta\hat{a}(\beta)J(S(t)\pi), & t > 0, \\ \mathrm{e}^{\beta t}, & t \leq 0. \end{cases}$$

This solution is real analytic in $(0,\infty)$ and belongs to $C^{0,\zeta}[0,T]$. As the definition of ψ_2 implies that $\psi_2 = \beta\hat{a}(\beta)\pi \otimes x$, we have $\mathcal{J}(\mathcal{S}(t)\psi_2) = \beta\hat{a}(\beta)J(S(t)\pi)x$ for every $t \geq 0$. Therefore the function $\tilde{u}_2 : \mathbb{R} \to X$ given by

$$\tilde{u}_2(t) := \begin{cases} \mathcal{J}(\mathcal{S}(t)\psi_2), & t \geq 0, \\ u_2(t), & t \leq 0, \end{cases}$$

is the unique solution to problem (4.2) with u_0 replaced by u_2, and $\tilde{u}_2$ is real analytic in $(0,\infty)$ and belongs to $C^{0,\zeta}([0,T];X)$. Since $u_0 = u_1 + u_2$, $\psi_0 = \psi_1 + \psi_2$, and $\zeta < 1 - \eta$, the conclusion of the theorem follows. □

Corollary 4.1.21 *Let u be the unique solution to problem (4.2) under the conditions stated in Theorem 4.1.20. Then the following holds:*

$$\big(\mathcal{S}(t)\psi_0\big)(\kappa) = \int_0^\infty \mathrm{e}^{-\kappa s} u(t-s)\,\mathrm{d}s, \quad t \geq 0,\ \kappa \geq 0,$$

where ψ_0 is given by (4.4).

4.1.2 The inhomogeneous problem

We consider the inhomogeneous Volterra equation

$$\begin{aligned} \frac{\mathrm{d}}{\mathrm{d}t}\int_0^t a(t-s)u(s)\,\mathrm{d}s &= f(t), \quad t > 0, \\ u(t) &= 0, \qquad t \leq 0, \end{aligned} \tag{4.16}$$

where the kernel $a : (0,\infty) \to \mathbb{R}$ belongs to the class $\mathcal{K}$ and where the function $f : [0,\infty) \to X$ belongs to $L^1(0,T;X)$ for every $T > 0$. Similar to the scalar inhomogeneous Volterra equation in Chapter 3 we shall consider problem (4.16) in an abstract setting. To this end let the Hilbert space $\mathcal{H}_{-1}$ be the completion of $\mathcal{H}$ with respect to the norm $\|\cdot\|_{\mathcal{H}_{-1}}$ given by

$$\|g\|_{\mathcal{H}_{-1}} := \big\|(\beta I_{\mathcal{H}} - \mathcal{A})^{-1} g\big\|_{\mathcal{H}}, \quad g \in \mathcal{H},$$

and let $\mathcal{A}_{-1} : \mathcal{H} \subseteq \mathcal{H}_{-1} \to \mathcal{H}_{-1}$ be the linear extension of $\mathcal{A}$ such that $\mathcal{A}_{-1}$ is the infinitesimal generator of the analytic semigroup $\{\mathcal{S}_{-1}(t)\}_{t\geq 0}$ on $\mathcal{H}_{-1}$, which denotes the continuous extension of $\{\mathcal{S}(t)\}_{t\geq 0}$.
We define the linear operators $\bar{\mathcal{A}} : \mathrm{D}(\mathcal{J}) \to \mathcal{H}$ and $\mathcal{I} : \mathrm{D}(\mathcal{J}) \to X$ by respectively

$$(\bar{\mathcal{A}}g)(\kappa) := \mathcal{J}(g) - \kappa g(\kappa), \quad g \in \mathrm{D}(\mathcal{J}),\ \kappa \geq 0,$$

$$\mathcal{I}(g) := \int_{[0,\infty)} \big(\mathcal{J}(g) - \kappa g(\kappa)\big)\, \nu(\mathrm{d}\kappa), \quad g \in \mathrm{D}(\mathcal{J}).$$

Note that $\mathcal{I}$ is well-defined by Lemma 4.1.7, and that if g belongs to $\mathrm{D}(\mathcal{J})$, then g belongs to $\mathrm{D}(\mathcal{A})$ if and only if $\mathcal{I}(g) = 0$. Moreover, the definition of the linear operator $\bar{A} : \mathrm{D}(J) \to H$ and the linear functional $I : \mathrm{D}(J) \to \mathbb{R}$ in Section 3.7 imply that $\bar{\mathcal{A}}(\varphi \otimes x) = \bar{A}\varphi \otimes x$ and $\mathcal{I}(\varphi \otimes x) = I(\varphi)x$ for all $\varphi \in \mathrm{D}(J)$ and $x \in X$. By Lemma 3.5.5 we have that the function π given by (4.15) belongs to $\mathrm{D}(J)$ and that $I(\pi) = 1$ and $\bar{A}\pi = \beta\pi$. These facts imply the following lemma.

Lemma 4.1.22 *The function $g - \pi \otimes \mathcal{I}(g)$ belongs to $\mathrm{D}(\mathcal{A})$ for every $g \in \mathrm{D}(\mathcal{J})$.*

Heuristically, our semigroup approach suggests that the function $\psi : [0,\infty) \times [0,\infty) \to X$ defined by

$$\psi(t,\kappa) := \begin{cases} \int_0^t e^{-\kappa(t-s)} u(s)\,\mathrm{d}s, & t > 0,\ \kappa \geq 0, \\ 0, & t = 0,\ \kappa \geq 0, \end{cases}$$

is such that $\psi(t,\cdot)$ belongs to $\mathrm{D}(\mathcal{J})$ with $\mathcal{J}(\psi(t,\cdot)) = u(t)$ and $\mathcal{I}(\psi(t,\cdot)) = f(t)$ for every $t > 0$, and

$$\begin{aligned} \tfrac{\mathrm{d}}{\mathrm{d}t}\psi(t,\cdot) &= \bar{\mathcal{A}}\psi(t,\cdot) = \mathcal{A}\big(\psi(t,\cdot) - \pi \otimes f(t)\big) + \bar{\mathcal{A}}(\pi \otimes f(t)) \\ &= \mathcal{A}_{-1}\psi(t,\cdot) - \mathcal{A}_{-1}(\pi \otimes f(t)) + \beta\pi \otimes f(t) \\ &= \mathcal{A}_{-1}\psi(t,\cdot) + (\beta I_H - A_{-1})\pi \otimes f(t), \quad t > 0. \end{aligned}$$

Therefore the abstract setting for problem (4.16) is the inhomogeneous abstract Cauchy problem

$$\begin{aligned} &\tfrac{\mathrm{d}}{\mathrm{d}t}\psi(t) = \mathcal{A}_{-1}\psi(t) + (\beta I_H - A_{-1})\pi \otimes f(t), \quad t > 0, \\ &\psi(0) = 0. \end{aligned} \tag{4.17}$$

Definition 4.1.23 The *mild solution* to problem (4.17) is the continuous function $\psi : [0,\infty) \to \mathcal{H}_{-1}$ defined by

$$\psi(t) := \int_0^t \big((\beta I_H - A)S(t-s)\pi \otimes f(s)\big)\,\mathrm{d}s, \quad t \geq 0.$$

The integral above is Bochner in $(\mathcal{H}, \mathrm{D}(\mathcal{A}))_{\theta,2}$ and hence, the mild solution ψ has values in $(\mathcal{H}, \mathrm{D}(\mathcal{A}))_{\theta,2}$ for all $t \geq 0$ and $\theta \in (0,1)$ such that $\theta < \frac{1+\alpha(a)}{2}$. Indeed, we observe that π belongs to $(H, \mathrm{D}(A))_{\theta,2}$ and $(\beta I_H - A)S(t)$ belongs to $\mathcal{L}\big((H, \mathrm{D}(A))_{\theta,2}\big)$ by Proposition 3.5.7 respectively Corollary 1.4.30(i). Compare also Proposition 3.7.2.

Proposition 4.1.24 *Let $f : [0, \infty) \to X$ belong to $L^p(0,T;X)$ for some $p \in [1, \infty)$ and every $T > 0$. Let ψ be the mild solution to problem (4.17). Then for all $T > 0$ and $\theta, \eta \in (0,1)$ such that $\eta < \theta < \frac{1+\alpha(a)}{2}$ the following holds:*

(i) *If $p = 1$, then ψ belongs to $L^q\big(0,T;(\mathcal{H}, \mathrm{D}(\mathcal{A}))_{\eta,2}\big)$ for every $q \in \big[1, \frac{1}{1-(\theta-\eta)}\big)$;*

(ii) *If $p \in \big(1, \frac{1}{\theta-\eta}\big)$, then ψ belongs to $L^q\big(0,T;(\mathcal{H}, \mathrm{D}(\mathcal{A}))_{\eta,2}\big)$ where $q := \frac{p}{1-(\theta-\eta)p}$;*

(iii) *If $p = \frac{1}{\theta-\eta}$, then ψ belongs to $L^q\big(0,T;(\mathcal{H}, \mathrm{D}(\mathcal{A}))_{\eta,2}\big)$ for every $q \in [1, \infty)$;*

(iv) *If $p \in \big(\frac{1}{\theta-\eta}, \infty\big)$, then ψ belongs to $C_0^{0,\theta-\eta-\frac{1}{p}}\big([0,T];(\mathcal{H}, \mathrm{D}(\mathcal{A}))_{\eta,2}\big)$.*

In particular, if f belongs to $C_0^{0,\gamma}([0,T];X)$ for some $\gamma \in (0,1)$ and every $T > 0$, then ψ belongs to

$$C^{[\gamma+\theta-\eta],\gamma+\theta-\eta-[\gamma+\theta-\eta]}\big([0,T];(\mathcal{H}, \mathrm{D}(\mathcal{A}))_{\eta,2}\big)$$

for all $T > 0$ and $\theta, \eta \in (0,1)$ such that $\eta < \theta < \frac{1+\alpha(a)}{2}$ and $\theta - \eta \neq 1 - \gamma$.

PROOF: The first part of the proposition is an application of Proposition 1.4.31(iv)-(vii). For the second part let $\theta \in (0,1)$ be such that $\theta < \frac{1+\alpha(a)}{2}$. Then π belongs to $(H, \mathrm{D}(A))_{\theta,2}$ and $\beta I_{\mathcal{H}} - \mathcal{A}_{-1}$ belongs to $\mathcal{L}\big((\mathcal{H}, \mathrm{D}(\mathcal{A}))_{\theta,2}, (\mathcal{H}_{-1}, \mathcal{H})_{\theta,2}\big)$ by Proposition 3.5.7 respectively 1.4.18, so that the mapping $t \mapsto (\beta I_H - A_{-1})\pi \otimes f(t)$ belongs to $C_0^{0,\gamma}\big([0,T];(\mathcal{H}_{-1}, \mathcal{H})_{\theta,2}\big)$ for every $T > 0$. Now the proposition is proved analogously to Proposition 3.7.3. □

Before we return to problem (4.16) we need a lemma concerning the Laplace transform of the mild solution ψ to problem (4.17).

Lemma 4.1.25 *Let $f : [0, \infty) \to X$ belong to $L^p(0,T;X)$ for some $p \in [1, \infty)$ and every $T > 0$. Let ψ be the mild solution to problem (4.17). If there exists $\lambda_0 \geq \beta$ such that $\int_0^\infty \mathrm{e}^{-\lambda_0 t}\|f(t)\|\,\mathrm{d}t < \infty$, then the Laplace transform $\hat{\psi} : (\lambda_0, \infty) \to (\mathcal{H}, \mathrm{D}(\mathcal{A}))_{\theta,2}$ is well-defined for every $\theta \in (0,1)$ such that $\theta < \frac{1+\alpha(a)}{2}$, and*

$$\hat{\psi}(\lambda) = (\lambda I_{H_{-1}} - A_{-1})^{-1}(\beta I_H - A_{-1})\pi \otimes \hat{f}(\lambda), \quad \lambda > \lambda_0.$$

Moreover, if a belongs to $\mathcal{K}$ with $\alpha(a) \in (0,1)$, then

$$\mathcal{J}(\hat{\psi}(\lambda)) = \frac{1}{\lambda\hat{a}(\lambda)}\hat{f}(\lambda), \quad \lambda > \lambda_0.$$

PROOF: Let $\lambda > \lambda_0$ and $\theta \in (0,1)$ such that $\theta < \frac{1+\alpha(a)}{2}$. Then existence of $\int_0^\infty e^{-\lambda t}\psi(t)\,\mathrm{d}t$ as a Bochner integral in $(\mathcal{H}, \mathrm{D}(\mathcal{A}))_{\theta,2}$ is proved in the same way as Lemma 3.7.4. Now we use Fubini's theorem and Proposition 1.4.6 to obtain

$$\begin{aligned}
\hat{\psi}(\lambda) &= \int_0^\infty \mathrm{e}^{-\lambda t}\left(\int_0^t ((\beta I_H - A)S(t-s)\pi \otimes f(s))\,\mathrm{d}s\right)\mathrm{d}t \\
&= \int_0^\infty \left(\mathrm{e}^{-\lambda s}\left(\int_s^\infty \mathrm{e}^{-\lambda(t-s)}(\beta I_H - A)S(t-s)\pi\,\mathrm{d}t\right) \otimes f(s)\right)\mathrm{d}s \\
&= \left(\int_0^\infty \mathrm{e}^{-\lambda t}S_{-1}(t)(\beta I_H - A_{-1})\pi\,\mathrm{d}t\right) \otimes \left(\int_0^\infty \mathrm{e}^{-\lambda s} f(s)\,\mathrm{d}s\right) \\
&= (\lambda I_{H_{-1}} - A_{-1})^{-1}(\beta I_H - A_{-1})\pi \otimes \hat{f}(\lambda).
\end{aligned}$$

The rest of the lemma is proved in the same way as Lemma 3.7.5. □

Now we are in a position to solve problem (4.16).

Theorem 4.1.26 *Let $f : [0,\infty) \to X$ belong to $L^p(0,T;X)$ for some $p \in [1,\infty)$ and every $T > 0$. If a belongs to $\mathcal{K}$ with $\alpha(a) \in (0,1)$, then problem (4.16) admits a unique solution u, and u has the representation*

$$u(t) = \begin{cases} \int_0^t J\big((\beta I_H - A)S(t-s)\pi\big) f(s)\,\mathrm{d}s, & t > 0, \\ 0, & t \le 0. \end{cases} \tag{4.18}$$

Moreover, for all $T > 0$ and $\zeta \in (0, \alpha(a))$ the following holds:

(i) *If $p = 1$, then u belongs to $L^q(0,T;X)$ for every $q \in \left[1, \frac{1}{1-\zeta}\right)$;*

(ii) *If $p \in \left(1, \frac{1}{\zeta}\right)$, then u belongs to $L^q(0,T;X)$ where $q := \frac{p}{1-\zeta p}$;*

(iii) *If $p = \frac{1}{\zeta}$, then u belongs to $L^q(0,T;X)$ for every $q \in [1,\infty)$;*

(iv) *If $p \in \left(\frac{1}{\zeta}, \infty\right)$, then u belongs to $C_0^{0,\zeta-\frac{1}{p}}([0,T];X)$.*

In particular, if f belongs to $C_0^{0,\gamma}([0,T];X)$ for some $\gamma \in (0,1)$ and every $T > 0$, then u belongs to

$$C^{[\gamma+\zeta],\gamma+\zeta-[\gamma+\zeta]}([0,T];X)$$

for all $T > 0$ and $\zeta \in (0,\alpha(a))$ such that $\zeta \neq 1-\gamma$.

PROOF: Let $T > 0$ and $\zeta \in (0, \alpha(a))$, and let $\eta := \frac{1-\zeta}{2}$ and $\theta := \frac{1+\zeta}{2}$. Then we have $\theta - \eta = \zeta$ and $0 < \frac{1-\alpha(a)}{2} < \eta \leq \theta < \frac{1+\alpha(a)}{2} < 1$. Let ψ be the mild solution to problem (4.17). Since $\psi(t)$ has values in $(\mathcal{H}, \mathrm{D}(\mathcal{A}))_{\eta,2}$ and $\mathcal{J}$ is continuous on $(\mathcal{H}, \mathrm{D}(\mathcal{A}))_{\eta,2}$ by Proposition 4.1.18, it follows that u given by (4.18) is such that $u(t) = \mathcal{J}(\psi(t))$ for every $t > 0$. Even more, the regularity properties of ψ in Proposition 4.1.24 imply those of u. □

4.2 Semilinear Volterra equations

We consider the semilinear X-valued Volterra integrodifferential equation

$$\begin{aligned} \frac{\mathrm{d}}{\mathrm{d}t} \int_{-\infty}^{t} a(t-s)u(s)\,\mathrm{d}s &= f(t, u(t)), \quad t > 0, \\ u(t) &= u_0(t), \qquad\quad t \leq 0, \end{aligned} \tag{4.19}$$

where the kernel $a : (0, \infty) \to \mathbb{R}$ belongs to the class $\mathcal{K}$ and where the functions $u_0 : (-\infty, 0] \to X$ respectively $f : [0, \infty) \times X \to X$ satisfy the following hypotheses:

Hypothesis 4.2.1 The function $u_0 : (-\infty, 0] \to X$ is strongly Borel measurable and has the following properties:

(i) There exist $M_1 > 0$ and $\omega > 0$ such that $\|u_0(t)\| \leq M_1 \mathrm{e}^{\omega t}$ for every $t \leq 0$;

(ii) There exist $M_2 > 0$ and $\delta > 0$ such that $\|u_0(0) - u_0(t)\| \leq M_2 |t|$ for every $t \in [-\delta, 0]$.

Hypothesis 4.2.2 The function $f : [0, \infty) \times X \to X$ is such that

(i) For every $T > 0$ there exists $L > 0$ such that

$$\big\|f(t, x) - f(t, y)\big\| \leq L\|x - y\|, \quad t \in [0, T],\ x, y \in X;$$

(ii) There exists $\vartheta \in (0, 1)$ such that for all $T > 0$ and $x_0 \in X$ there exist $r > 0$ and $K > 0$ such that

$$\big\|f(t, x) - f(s, x)\big\| \leq K|t - s|^{\vartheta}, \quad s, t \in [0, T],\ x \in B_X(x_0; r).$$

We consider the following notion of solution:

Definition 4.2.3 A solution to problem (4.19) is a strongly Borel measurable function $u : \mathbb{R} \to X$ with the following properties:

(i) $\int_{-\infty}^{t} a(t-s)\|u(s)\|\,\mathrm{d}s < \infty$ for every $t \geq 0$;

(ii) The mapping $t \mapsto \int_{-\infty}^{t} a(t-s)u(s)\,\mathrm{d}s$ belongs to $W^{1,1}([0,T];X)$ for every $T > 0$;

(iii) u satisfies (4.19) for almost every $t \in \mathbb{R}$.

Analogous to Section 3.1 we have that a solution u to problem (4.19) always belongs to $L^1(0,T;X)$ for every $T > 0$. We also have uniqueness of solutions as shown in the next theorem:

Theorem 4.2.4 *Problem (4.19) admits at most one solution.*

PROOF: We assume that u_1, $u_2 : \mathbb{R} \to X$ both are solutions to problem (4.19) and define $u : \mathbb{R} \to X$ by

$$u(t) := u_1(t) - u_2(t), \quad t \in \mathbb{R}.$$

By Definition 4.2.3(iii) we have $u(t) = 0$ for almost every $t \leq 0$, so we can define

$$T_0 := \sup\big\{T \geq 0;\ u(t) = 0 \text{ for almost every } t \leq T\big\}.$$

To prove the theorem it is sufficient to show that $T_0 = +\infty$. Seeking a contradiction, we suppose that $T_0 < \infty$. Definition 4.2.3(ii) and (iii) imply that

$$\int_0^t a(t-s)u(s)\,\mathrm{d}s = \int_0^t \big(f(s,u_1(s)) - f(s,u_2(s))\big)\,\mathrm{d}s, \quad t \geq 0. \tag{4.20}$$

Let $b \in \mathcal{K}$ denote the resolvent of the first kind of a, thus b satisfies

$$\int_0^t b(t-s)a(s)\,\mathrm{d}s = 1, \quad t > 0, \tag{4.21}$$

Using (4.21) and Fubini's theorem we have

$$\begin{aligned}\int_0^t u(s)\,\mathrm{d}s &= \int_0^t \left(\int_0^{t-s} b(t-s-\sigma)a(\sigma)\,\mathrm{d}\sigma\right) u(s)\,\mathrm{d}s \\ &= \int_0^t b(t-\sigma)\left(\int_0^{\sigma} a(\sigma-s)u(s)\,\mathrm{d}s\right)\mathrm{d}\sigma, \qquad t > 0.\end{aligned}$$

Combined with (4.20) it follows that for every $t > 0$,

$$\int_0^t u(s)\,\mathrm{d}s = \int_0^t b(t-\sigma)\left(\int_0^{\sigma} \big(f(s,u_1(s)) - f(s,u_2(s))\big)\,\mathrm{d}s\right)\mathrm{d}\sigma. \tag{4.22}$$

It is a result of Hypothesis 4.2.2(i) that for every $T > 0$ there exists $L > 0$ such that

$$\int_0^t b(t-\sigma)\left(\int_0^\sigma \big\|f(s,u_1(s)) - f(s,u_2(s))\big\| \,\mathrm{d}s\right) \mathrm{d}\sigma \\ \leq L\left(\int_0^t \|u(s)\| \,\mathrm{d}s\right)\cdot\left(\int_0^t b(\sigma)\,\mathrm{d}\sigma\right) < \infty, \qquad t \in (0,T].$$

Hence, we can apply Fubini's theorem in (4.22) to obtain

$$\int_0^t u(s)\,\mathrm{d}s = \int_0^t \left(\int_0^s b(s-\sigma)\big(f(\sigma,u_1(\sigma)) - f(\sigma,u_2(\sigma))\big)\,\mathrm{d}\sigma\right)\mathrm{d}s, \quad t > 0.$$

It follows that for almost every $t > 0$,

$$u(t) = \int_0^t b(t-s)\big(f(s,u_1(s)) - f(s,u_2(s))\big)\,\mathrm{d}s, \tag{4.23}$$

see [Tay65, page 415, Lemma 9-8V]. Now we fix an arbitrary $T_1 > T_0$. By Hypothesis 4.2.2(i) there exists $L_1 > 0$ such that

$$\big\|f(t,x) - f(t,y)\big\| \leq L_1\|x-y\|, \quad t \in [0,T_1],\ x,y \in X. \tag{4.24}$$

Since $\lim_{t\downarrow 0}\int_0^t b(s)\,\mathrm{d}s = 0$ there exists $M \in (0, T_1 - T_0)$ such that $\int_0^M b(s)\,\mathrm{d}s < \frac{1}{L_1}$. Let $T := T_0 + M$. Using (4.23), (4.24), the definition of T_0 and T, and Fubini's theorem, we have

$$\int_{T_0}^T \|u(t)\|\,\mathrm{d}t \leq \int_{T_0}^T \left(\int_0^t b(t-s)\|f(s,u_1(s)) - f(s,u_2(s))\|\,\mathrm{d}s\right)\mathrm{d}t \\ \leq L_1\int_{T_0}^T \left(\int_0^t b(t-s)\|u(s)\|\,\mathrm{d}s\right)\mathrm{d}t = L_1\int_{T_0}^T \left(\int_{T_0}^t b(t-s)\|u(s)\|\,\mathrm{d}s\right)\mathrm{d}t \\ = L_1\int_{T_0}^T \|u(s)\|\left(\int_0^{T-s} b(t)\,\mathrm{d}t\right)\mathrm{d}s \leq \underbrace{\left(L_1\int_0^M b(t)\,\mathrm{d}t\right)}_{<1}\cdot\int_{T_0}^T \|u(s)\|\,\mathrm{d}s.$$

This implies that $\int_{T_0}^T \|u(s)\|\,\mathrm{d}s = 0$. It follows that $u(t) = 0$ for almost every $t \in [T_0, T]$ and hence, for almost every $t \leq T$. This contradicts the definition of T_0 and thus we have $T_0 = +\infty$. □

Following our semigroup approach as in Section 4.1 we consider problem (4.19) in an abstract setting, that is, the semilinear initial value problem

$$\begin{aligned} \tfrac{\mathrm{d}}{\mathrm{d}t}\psi(t) &= \mathcal{A}_{-1}\psi(t) + (\beta I_H - A_{-1})\pi \otimes f\big(t, \mathcal{J}(\psi(t))\big), \quad t > 0, \\ \psi(0) &= \psi_0, \end{aligned} \tag{4.25}$$

where $\psi_0 : [0,\infty) \to X$ is defined by

$$\psi_0(\kappa) := \int_0^\infty \mathrm{e}^{-\kappa t} u_0(-t)\,\mathrm{d}t, \quad \kappa \geq 0. \tag{4.26}$$

We solve this initial value problem in Section 4.2.1, to return to the Volterra equation in Section 4.2.2.

4.2.1 The semilinear initial value problem

We start with some general theory concerning semilinear initial value problems. This will ease the proof when solving our specific semilinear initial value problem (4.25).

Let $(E, \|\cdot\|_E)$ be a Banach space and $\{T(t)\}_{t\geq 0}$ an analytic semigroup on E with infinitesimal generator $B : \mathrm{D}(B) \subseteq E \to E$. For every $\alpha \in (0,1)$ we denote

$$(E_\alpha, \|\cdot\|_\alpha) := \big((E, \mathrm{D}(B))_{\alpha,2}, \|\cdot\|_{(E,\mathrm{D}(B))_{\alpha,2}}\big).$$

For all $\alpha \in (0,1)$, $x_0 \in E_\alpha$, and $r > 0$ we denote

$$B_\alpha(x_0; r) := \big\{x \in E_\alpha;\ \|x - x_0\|_\alpha \leq r\big\}.$$

We consider the semilinear initial value problem

$$\begin{aligned} \tfrac{\mathrm{d}}{\mathrm{d}t} v(t) &= Bv(t) + F(t, v(t)), \quad t > 0, \\ v(0) &= v_0, \end{aligned} \tag{4.27}$$

where v_0 belongs to E and where the function $F : [0,\infty) \times E_\alpha \to E$ is continuous for some $\alpha \in (0,1)$.

Definition 4.2.5 A *strict solution* to problem (4.27) is a function $v : [0,\infty) \to E$ such that v belongs to $C\big([0,\infty); \mathrm{D}(B)\big) \cap C^1\big([0,\infty); E\big)$, where $\mathrm{D}(B)$ is endowed with the graph norm of B, and such that v satisfies (4.27) for every $t \geq 0$.

Hypothesis 4.2.6 The function $F : [0,\infty) \times E_\alpha \to E$ is such that

(i) For all $T > 0$ and $R > 0$ there exists $L > 0$ such that

$$\big\|F(t,x) - F(t,y)\big\|_E \leq L\|x - y\|_\alpha, \quad t \in [0,T],\ x, y \in B_\alpha(0; R);$$

(ii) There exists $\vartheta \in (0,1)$ such that for all $T > 0$ and $v_0 \in E_\alpha$ there exist $r > 0$ and $K > 0$ such that

$$\big\|F(t,x) - F(s,x)\big\|_E \leq K|t - s|^\vartheta, \quad s, t \in [0,T],\ x \in B_\alpha(v_0; r).$$

Theorem 4.2.7 *Let F satisfy Hypothesis 4.2.6 for some $\alpha \in (0,1)$. If v_0 belongs to $\mathrm{D}(B)$ and $Bv_0 + F(0, v_0)$ belongs to $\overline{\mathrm{D}(B)}$, then problem (4.27) admits a unique strict solution v, and v satisfies*

$$v(t) = T(t)v_0 + \int_0^t T(t-s)F(s, v(s))\,\mathrm{d}s, \quad t \geq 0.$$

For a proof of Theorem 4.2.7 see [Lun95, page 268, Proposition 7.1.10(iii)]. Now we solve our semilinear initial value problem (4.25).

Theorem 4.2.8 *Let u_0 and f satisfy Hypothesis 4.2.1 respectively 4.2.2. If a belongs to $\mathcal{K}$ with $\alpha(a) \in (0,1)$, then problem (4.25) admits a unique strict solution $\psi : [0,\infty) \to (\mathcal{H}, \mathrm{D}(\mathcal{A}))_{\theta,2}$ for every $\theta \in (0,1)$ such that $\theta < \frac{1+\alpha(a)}{2}$. Moreover, ψ belongs to $C^{0,\theta-\eta}\big([0,T]; (\mathcal{H}, \mathrm{D}(\mathcal{A}))_{\eta,2}\big)$ for all $T > 0$ and $\theta, \eta \in (0,1)$ such that $\frac{1-\alpha(a)}{2} < \eta < \theta$, and ψ satisfies*

$$\psi(t) = \mathcal{S}(t)\psi_0 + \int_0^t \Big((\beta I_H - A)S(t-s)\pi \otimes f\big(s, \mathcal{J}(\psi(s))\big)\Big)\,\mathrm{d}s, \quad t \geq 0,$$

where the integral is Bochner in $(\mathcal{H}, \mathrm{D}(\mathcal{A}))_{\eta,2}$.

PROOF: Let $\theta, \eta \in (0,1)$ be such that $\frac{1-\alpha(a)}{2} < \eta < \theta < \frac{1+\alpha(a)}{2}$. Note that $(\beta I_H - A_{-1})\pi$ belongs to $(H_{-1}, H)_{\theta,2}$ by Propositions 3.5.7 and 1.4.18. Moreover, ψ_0 belongs to $(\mathcal{H}, \mathrm{D}(\mathcal{A}))_{\theta,2}$ as a result of the proof of Theorem 4.1.20. We shall apply Theorem 4.2.7 with

$$E := (\mathcal{H}_{-1}, \mathcal{H})_{\theta,2},$$

and $B : \mathrm{D}(B) \subseteq E \to E$ the part of $\mathcal{A}_{-1}$ in E. Note that by Lemma 1.4.19 and Proposition 1.4.18 we have that B is the infinitesimal generator of the analytic semigroup $\{\mathcal{S}_{-1}(t)|_E\}_{t\geq 0}$ on E, and that

$$\mathrm{D}(B) = (\mathcal{H}, \mathrm{D}(\mathcal{A}))_{\theta,2}.$$

Furthermore, we take $\alpha := 1 + \eta - \theta$. Then it follows from Corollary 1.4.23 and Theorem 1.4.21 that

$$\begin{aligned} E_\alpha &= (E, \mathrm{D}(B))_{\alpha,2} = \big((\mathcal{H}_{-1}, \mathcal{H})_{\theta,2}, (\mathcal{H}, \mathrm{D}(\mathcal{A}))_{\theta,2}\big)_{\alpha,2} \\ &= \big((\mathcal{H}_{-1}, \mathrm{D}(\mathcal{A}))_{\frac{1}{2}\theta,2}, (\mathcal{H}_{-1}, \mathrm{D}(\mathcal{A}))_{\frac{1}{2}(\theta+1),2}\big)_{\alpha,2} \\ &= (\mathcal{H}_{-1}, \mathrm{D}(\mathcal{A}))_{\frac{1}{2}(\theta+\alpha),2} = (\mathcal{H}, \mathrm{D}(\mathcal{A}))_{\theta+\alpha-1,2} = (\mathcal{H}, \mathrm{D}(\mathcal{A}))_{\eta,2}. \end{aligned}$$

Now we define $F : [0,\infty) \times E_\alpha \to E$ by

$$F(t,x) := (\beta I_H - A_{-1})\pi \otimes f\big(t, \mathcal{J}(x)\big), \quad t \geq 0, \, x \in E_\alpha,$$

which is well-defined since $\mathcal{J}$ belongs to $\mathcal{L}(E_\alpha; X)$ by Proposition 4.1.18. We show that F satisfies Hypothesis 4.2.6, omitting the subscripts of $\|\mathcal{J}\|_{\mathcal{L}(E_\alpha;X)}$ and $\|(\beta I_H - A_{-1})\pi\|_{(H_{-1},H)_{\theta,2}}$. For Hypothesis 4.2.6(i) let $T, R > 0$, $t \in [0,T]$, and $x, y \in B_\alpha(0; R)$. By Hypothesis 4.2.2(i) there exists $L > 0$ independent of t, x, and y, such that

$$\begin{aligned}\big\|F(t,x) - F(t,y)\big\|_E &= \big\|(\beta I_H - A_{-1})\pi\big\| \, \big\|f\big(t, \mathcal{J}(x)\big) - f\big(t, \mathcal{J}(y)\big)\big\| \\ &\le L\big\|(\beta - A_{-1})\pi\big\| \, \|\mathcal{J}\| \, \|x - y\|_\alpha.\end{aligned}$$

For Hypothesis 4.2.6(ii) let $T > 0$, $s, t \in [0,T]$, and $v_0 \in E_\alpha$. By Hypothesis 4.2.2(ii) with x_0 chosen as $\mathcal{J}(v_0)$ there exists $\vartheta \in (0,1)$ independent of s, t, and v_0, and there exist $r, K > 0$, both independent of s and t, such that for every $x \in B_\alpha\big(v_0; \frac{r}{\|\mathcal{J}\|}\big)$ we have that $\mathcal{J}(x)$ belongs to $B_X(x_0; r)$ and

$$\begin{aligned}\big\|F(t,x) - F(s,x)\big\|_{E_\alpha} &= \big\|(\beta I_H - A_{-1})\pi\big\| \, \big\|f\big(t, \mathcal{J}(x)\big) - f\big(s, \mathcal{J}(x)\big)\big\| \\ &\le K\big\|(\beta I_H - A_{-1})\pi\big\| \, |t - s|^\vartheta.\end{aligned}$$

Hence, F satisfies Hypothesis 4.2.6 and we are in a position to apply Theorem 4.2.7 with v_0 replaced by ψ_0. We obtain that problem (4.25) admits a unique strict solution ψ belonging to $C\big([0,\infty); \mathrm{D}(B)\big) \cap C^1\big([0,\infty); E\big)$. In particular, ψ belongs to $C^{0,1-\alpha}\big([0,T]; E_\alpha\big)$ for every $T > 0$ by Proposition 1.4.20. Moreover, ψ satisfies

$$\begin{aligned}\psi(t) &= \mathcal{S}_{-1}(t)\psi_0 + \int_0^t \mathcal{S}_{-1}(t-s)F(s, \psi(s)) \, \mathrm{d}s \\ &= \mathcal{S}(t)\psi_0 + \int_0^t \Big(\big(\beta I_H - A\big)S(t-s)\pi \otimes f\big(s, \mathcal{J}(\psi(s))\big)\Big) \, \mathrm{d}s, \qquad t \ge 0,\end{aligned}$$

where the integrals are Bochner in E_α. □

4.2.2 The semilinear Volterra equation

In this section we solve the semilinear Volterra equation (4.19).

Theorem 4.2.9 *Let u_0 and f satisfy Hypothesis 4.2.1 respectively 4.2.2. If a belongs to $\mathcal{K}$ with $\alpha(a) \in (0,1)$, then problem (4.19) admits a unique solution u, and u has the representation*

$$u(t) = \begin{cases} \mathcal{J}\big(\psi(t)\big), & t > 0, \\ u_0(t), & t \le 0, \end{cases} \tag{4.28}$$

where ψ is the strict solution to problem (4.25). Furthermore, u belongs to $C^{0,\zeta}([0,T];X)$ for all $T>0$ and $\zeta \in (0,\alpha(a))$, and u satisfies

$$u(t) = \mathcal{J}\big(\mathcal{S}(t)\psi_0\big) + \int_0^t J\big((\beta I_H - A)S(t-s)\pi\big) f(s,u(s))\,\mathrm{d}s, \quad t \geq 0, \tag{4.29}$$

where ψ_0 is given by (4.26).

PROOF: Let $T>0$ and $\zeta \in (0,\alpha(a))$, and let $\eta := \frac{1-\zeta}{2}$ and $\theta := \frac{1+\zeta}{2}$. Then we have $\theta - \eta = \zeta$ and $\frac{1-\alpha(a)}{2} < \eta < \theta < \frac{1+\alpha(a)}{2}$. Theorem 4.2.8 implies that ψ belongs to $C^{\zeta}\big([0,T];(\mathcal{H},\mathrm{D}(\mathcal{A}))_{\eta,2}\big)$ and satisfies

$$\psi(t) = \mathcal{S}(t)\psi_0 + \int_0^t \Big(\big(\beta I_H - A\big)S(t-s)\pi \otimes f\big(s,\mathcal{J}(\psi(s))\big)\Big)\,\mathrm{d}s, \quad t \geq 0.$$

Let $\psi_1,\psi_2 : [0,\infty) \to (\mathcal{H},\mathrm{D}(\mathcal{A}))_{\eta,2}$ be defined by respectively

$$\psi_1(t) := \mathcal{S}(t)\psi_0, \quad t \geq 0,$$

$$\psi_2(t) := \int_0^t \Big(\big(\beta I_H - A\big)S(t-s)\pi \otimes f\big(s,\mathcal{J}(\psi(s))\big)\Big)\,\mathrm{d}s, \quad t \geq 0.$$

Note that $\psi = \psi_1 + \psi_2$. Let $\tilde{f} : [0,\infty) \to X$ be defined by

$$\tilde{f}(t) := f\big(t,\mathcal{J}(\psi(t))\big), \quad t \geq 0.$$

Note that $\tilde{f}$ is continuous, since f is continuous by Hypothesis 4.2.2, $\mathcal{J}$ is continuous on $(\mathcal{H},\mathrm{D}(\mathcal{A}))_{\eta,2}$ by Proposition 4.1.18, and ψ is continuous with values in $(\mathcal{H},\mathrm{D}(\mathcal{A}))_{\eta,2}$. It is a result of Theorem 4.1.20 that problem (4.19) with f identically zero admits a unique solution u_1, and u_1 belongs to $C^{0,\zeta}([0,T];X)$ and has the representation

$$u_1(t) = \begin{cases} \mathcal{J}\big(\mathcal{S}(t)\psi_0\big), & t > 0, \\ u_0(t), & t \leq 0. \end{cases}$$

Moreover, Theorem 4.1.26 implies that problem (4.19) with u_0 identically zero and $f(t,u(t))$ replaced by $\tilde{f}(t)$ admits a unique solution u_2 and u_2 belongs to $C_0^{0,\zeta}([0,T];X)$ and has the representation

$$u_2(t) = \begin{cases} \int_0^t J\big((\beta I_H - A)S(t-s)\pi\big)\tilde{f}(s)\,\mathrm{d}s, & t > 0, \\ 0, & t \leq 0. \end{cases}$$

It follows that the function $u := u_1 + u_2$ is the unique solution to problem (4.19) and u belongs to $C^{0,\zeta}([0,T];X)$. Since $u_1(t) = \mathcal{J}(\psi_1(t))$ and $u_2(t) = \mathcal{J}(\psi_2(t))$ for every $t > 0$, u has the representation (4.28) and hence, u satisfies (4.29). □

4.3 First kind linear Volterra equations

We consider the first kind linear X-valued Volterra equation

$$\int_0^t a(t-s)u(s)\,\mathrm{d}s = h(t), \quad t \geq 0, \tag{4.30}$$

where the kernel $a : (0,\infty) \to \mathbb{R}$ belongs to the class $\mathcal{K}$ and where the function $h : [0,\infty) \to X$ belongs to $C([0,\infty); X)$. We consider the following notion of solution:

Definition 4.3.1 A solution to problem (4.30) is a strongly Borel measurable function $u : [0,\infty) \to X$ such that

(i) $\int_0^t a(t-s)\|u(s)\|\,\mathrm{d}s < \infty$ for every $t > 0$;

(ii) u satisfies (4.30) for every $t \geq 0$.

We remark that a function $u : [0,\infty) \to X$ is strongly Borel measurable if and only if the mapping $t \mapsto \langle u(t), x\rangle$ is Borel measurable on $[0,\infty)$ for every $x \in X$. Moreover, analogous to Section 3.1 we have that a solution u to problem (4.30) always belongs to $L^1(0,T;X)$ for every $T > 0$. Concerning uniqueness of solutions we have the following result:

Theorem 4.3.2 *Problem (4.30) admits at most one solution.*

PROOF: Let u be a solution to problem (4.30) with h identically zero. Then Definition 4.3.1(ii) implies that

$$\int_0^t a(t-s)\langle u(s), x\rangle\,\mathrm{d}s = 0, \quad t \geq 0,\ x \in X.$$

It follows from Corollary 1.2.2 that $\langle u(t), x\rangle = 0$ for every $x \in X$ and almost every $t > 0$. Therefore, $u(t) = 0$ for almost every $t > 0$. □

Theorem 4.3.3 *Let h belong to $C_0^{0,\gamma}([0,T];X)$ for some $\gamma \in \big(1-\alpha(a), 1\big)$ and every $T > 0$. If a belongs to $\mathcal{K}$ with $\alpha(a) \in (0,1)$, then problem (4.30) admits a unique solution u. Moreover, u belongs to $C^{0,\gamma+\zeta-1}([0,T];X)$ for all $T > 0$ and $\zeta \in \big(1-\gamma, \alpha(a)\big)$, and u has the representation*

$$u(t) = \frac{\mathrm{d}}{\mathrm{d}t}\int_0^t J\big((\beta I_H - A)S(t-s)\pi\big)h(s)\,\mathrm{d}s, \quad t \geq 0. \tag{4.31}$$

PROOF: Let $\zeta \in \big(1-\gamma, \alpha(a)\big)$. It is a result of Theorem 4.1.26 that the linear X-valued Volterra integrodifferential equation

$$\frac{\mathrm{d}}{\mathrm{d}t}\int_0^t a(t-s)\tilde{u}(s)\,\mathrm{d}s = h(t), \quad t>0,$$
$$\tilde{u}(t) = 0, \qquad t \le 0,$$

admits a unique solution $\tilde{u}:\mathbb{R}\to X$ according to Definition 4.1.1. Moreover, $\tilde{u}$ belongs to $C^{1,\gamma+\zeta-1}([0,T];X)$ for every $T>0$ and has the representation

$$\tilde{u}(t) = \begin{cases} \int_0^t J\big((\beta I_H - A)S(t-s)\pi\big)h(s)\,\mathrm{d}s, & t>0,\\ 0, & t\le 0.\end{cases}$$

It follows that u given by (4.31) is well-defined and belongs to $C^{0,\gamma+\zeta-1}([0,T];X)$ for every $T>0$. Therefore u is strongly Borel measurable and satisfies Definition 4.3.1(i). Finally, u satisfies Definition 4.3.1(ii), since

$$\begin{aligned}\int_0^t a(t-s)u(s)\,\mathrm{d}s &= \int_0^t a(t-s)\big(\tfrac{\mathrm{d}}{\mathrm{d}s}\tilde{u}(s)\big)\,\mathrm{d}s\\ &= -a(t)\tilde{u}(0) + \frac{\mathrm{d}}{\mathrm{d}t}\int_0^t a(t-s)\tilde{u}(s)\,\mathrm{d}s = h(t), \quad t>0.\end{aligned}$$

□

Theorem 4.3.3 assumes that $h(0)=0$. However, this is not necessary as the next corollary shows:

Corollary 4.3.4 *Let h belong to $C^{0,\gamma}([0,T];X)$ for some $\gamma \in \big(1-\alpha(a),1\big)$ and every $T>0$. If a belongs to $\mathcal{K}$ with $\alpha(a)\in(0,1)$, then problem (4.30) admits a unique solution u, and u has the representation*

$$u(t) = b(t)h(0) + \frac{\mathrm{d}}{\mathrm{d}t}\int_0^t J\big((\beta I_H - A)S(t-s)\pi\big)\big(h(s)-h(0)\big)\,\mathrm{d}s, \quad t\ge 0,$$

where b denotes the resolvent of the first kind of a.

Chapter 5

Stochastic linear Volterra equations

In this chapter we consider a stochastic perturbation of the scalar linear Volterra integrodifferential equation. We start with some preliminaries in Section 5.1. In Section 5.2 we interpret the stochastic linear Volterra equation as a first kind Hilbert-valued Volterra equation. Then Section 5.3 is concerned with the representation of a solution to the stochastic Volterra equation by stochastic convolution.

5.1 Preliminaries

5.1.1 Gaussian measures

Definition 5.1.1 A *Gaussian measure on* $\mathbb{R}$ is a Borel probability measure μ on $\mathbb{R}$ such that either for some $m \in \mathbb{R}$, μ is the Dirac measure δ_m, or for some $m \in \mathbb{R}$ and $C > 0$,

$$\mu(B) = \frac{1}{\sqrt{2\pi C}} \int_B \mathrm{e}^{-\frac{(x-m)^2}{2C}}\, \mathrm{d}x, \quad B \in \mathcal{B}(\mathbb{R}).$$

The numbers m and C, with $C := 0$ in case of a Dirac measure, are called the *mean* respectively *variance* of μ. If $m = 0$, then μ is called a *centered* Gaussian measure.

Proposition 5.1.2 *If μ is a Gaussian measure on $\mathbb{R}$ with mean m and variance C, then the following holds:*

$$\int_{\mathbb{R}} |x-m|^r \, \mu(\mathrm{d}x) = \frac{(2C)^{\frac{r}{2}}}{\sqrt{\pi}} \Gamma\left(\frac{r+1}{2}\right), \quad r \geq 0,$$

$$\int_{\mathbb{R}} (x-m)^{2n} \, \mu(\mathrm{d}x) = C^n \cdot 1 \cdot 3 \cdot 5 \cdot \dots \cdot (2n-1), \quad n = 1, 2, \cdots,$$

$$\int_{\mathbb{R}} (x-m)^{2n-1}\, \mu(\mathrm{d}x) = 0, \quad n = 1, 2, \cdots .$$

In particular,

$$m = \int_{\mathbb{R}} x\, \mu(\mathrm{d}x), \qquad C = \int_{\mathbb{R}} (x-m)^2\, \mu(\mathrm{d}x).$$

Proposition 5.1.3 *A Borel probability measure μ on $\mathbb{R}$ is Gaussian with mean m and variance C if and only if*

$$\int_{\mathbb{R}} \mathrm{e}^{\mathrm{i}ux}\, \mu(\mathrm{d}x) = \mathrm{e}^{\mathrm{i}um - \frac{1}{2}Cu^2}, \quad u \in \mathbb{R}.$$

Definition 5.1.4 Let $d \in \mathbb{N}$. A *Gaussian measure on* $\mathbb{R}^d$ is a Borel probability measure μ on $\mathbb{R}^d$ such that for some $m \in \mathbb{R}^d$ and some symmetric, positive semidefinite matrix $C \in \mathbb{R}^{d\times d}$,

$$\int_{\mathbb{R}^d} \mathrm{e}^{\mathrm{i}\langle u,x\rangle}\, \mu(\mathrm{d}x) = \mathrm{e}^{\mathrm{i}\langle u,m\rangle - \frac{1}{2}\langle Cu,u\rangle}, \quad u \in \mathbb{R}^d.$$

The vector m and the matrix C are called the *mean* respectively *covariance* of μ. If $m = 0$, then μ is called a *centered* Gaussian measure.

Proposition 5.1.5 *Let $d \in \mathbb{N}$. If μ is a Gaussian measure on $\mathbb{R}^d$ with mean m and covariance C, then for all $j, k = 1, \cdots, d$,*

$$m_j = \int_{\mathbb{R}^d} x_j\, \mu(\mathrm{d}x), \qquad C_{j,k} = \int_{\mathbb{R}^d} (x_j - m_j)(x_k - m_k)\, \mu(\mathrm{d}x).$$

If, in addition, C is nonsingular, then

$$\mu(B) = \frac{1}{\sqrt{(2\pi)^d \det C}} \int_B \mathrm{e}^{-\frac{1}{2}\left\langle C^{-1}(x-m), x-m\right\rangle}\, \mu(\mathrm{d}x), \quad B \in \mathcal{B}(\mathbb{R}^d).$$

Proposition 5.1.6 *Let $d \in \mathbb{N}$. A Borel probability measure μ on $\mathbb{R}^d$ is Gaussian if and only if for every $\xi \in \mathbb{R}^d$ the image measure $\langle \xi, \mu\rangle$ on $\mathbb{R}$ given by*

$$\langle \xi, \mu\rangle(B) := \mu\{x \in \mathbb{R}^d;\ \langle \xi, x\rangle \in B\}, \quad B \in \mathcal{B}(\mathbb{R}),$$

is a Gaussian measure on $\mathbb{R}$.

Definition 5.1.7 Let E be a separable real Banach space. A *(centered) Gaussian measure on E* is a Borel probability measure μ on E such that for every $\varphi \in E'$ the image measure $\langle \varphi, \mu\rangle$ on $\mathbb{R}$ given by

$$\langle \varphi, \mu\rangle(B) := \mu\{x \in E;\ \varphi(x) \in B\}, \quad B \in \mathcal{B}(\mathbb{R}),$$

is a (centered) Gaussian measure on $\mathbb{R}$.

Proposition 5.1.8 *A Borel probability measure μ on a separable real Hilbert space $(H, \langle\cdot,\cdot\rangle, \|\cdot\|)$ is a Gaussian measure on H if and only if there exist a vector $m \in H$ and a nonnegative*[1]*, symmetric, trace class operator $C \in \mathcal{L}(H)$ such that*

$$\int_H \mathrm{e}^{\mathrm{i}\langle u,x\rangle}\,\mu(\mathrm{d}x) = \mathrm{e}^{\mathrm{i}\langle u,m\rangle - \frac{1}{2}\langle Cu,u\rangle}, \quad u \in H.$$

In that case

$$\int_H \langle u,x\rangle\,\mu(\mathrm{d}x) = \langle u,m\rangle, \quad u \in H,$$

$$\int_H \langle u_1,x\rangle\langle u_2,x\rangle\,\mu(\mathrm{d}x) - \langle u_1,m\rangle\langle u_2,m\rangle = \langle Cu_1,u_2\rangle, \quad u_1,u_2 \in H,$$

$$\operatorname{Tr} C = \int_H \|x-m\|^2\,\mu(\mathrm{d}x).$$

Definition 5.1.9 Let μ be a Gaussian measure on a separable real Hilbert space H. The vector $m \in H$ and the operator $C \in \mathcal{L}(H)$ with the properties stated in Proposition 5.1.8 are called the *mean* respectively *covariance* of μ.

Definition 5.1.10 Let $(\Omega, \mathcal{F}, \mathbb{P})$ be a probability space.

(i) A *(centered) Gaussian variable* is a random variable $X : \Omega \to \mathbb{R}$ such that its distribution is a (centered) Gaussian measure on $\mathbb{R}$;

(ii) A *(centered) Gaussian vector* on a separable real Banach space E is a random vector $X : \Omega \to E$ such that its distribution is a (centered) Gaussian measure on E;

(iii) A *(centered) Gaussian system* is a collection $\{X_\alpha\}_{\alpha\in\mathcal{I}}$, indexed by a non-empty index set $\mathcal{I}$, of random variables on Ω such that for all $n \in \mathbb{N}$ and $\alpha_1, \cdots, \alpha_n \in \mathcal{I}$, $(X_{\alpha_1}, \cdots, X_{\alpha_n})$ is a (centered) Gaussian vector on $\mathbb{R}^n$. It is said to have *mean* $\{m_\alpha\}_{\alpha\in\mathcal{I}}$ and *covariance* $\{C_{\alpha,\beta}\}_{\alpha,\beta\in\mathcal{I}}$ where for all $\alpha, \beta \in \mathcal{I}$,

$$m_\alpha := \mathbb{E}X_\alpha, \quad C_{\alpha,\beta} := \mathbb{E}\big((X_\alpha - m_\alpha)(X_\beta - m_\beta)\big).$$

Remark 5.1.11 Proposition 5.1.2 implies that if X is a Gaussian variable on a probability space $(\Omega, \mathcal{F}, \mathbb{P})$, then X belongs to $L^p(\Omega)$ for every $p \in [1, \infty)$.

[1]Nonnegative in the sense that for every $u \in H$, $\langle Cu, u\rangle \geq 0$.

Proposition 5.1.12 *Let $d \in \mathbb{N}$.*

(i) *A random vector $(X_1, \cdots, X_d)$ on $\mathbb{R}^d$ is Gaussian if and only if for all $\lambda_1, \cdots, \lambda_d \in \mathbb{R}$, $\sum_{j=1}^{d} \lambda_j X_j$ is a Gaussian variable;*

(ii) *If $(X_1, \cdots, X_d)$ is a Gaussian vector on $\mathbb{R}^d$, then $X_1, \cdots, X_d$ are independent if and only if for all $j, k = 1, \cdots, d$ such that $j \neq k$,*

$$\mathbb{E}\big((X_j - \mathbb{E}X_j)(X_k - \mathbb{E}X_k)\big) = 0;$$

(iii) *If $X_1, \cdots, X_d$ are independent Gaussian variables, then $(X_1, \cdots, X_d)$ is a Gaussian vector on $\mathbb{R}^d$.*

For the sentences 5.1.1-5.1.6 we refer to [Bog98, pages 1-4, Sections 1.1-1.2]. In particular, for Proposition 5.1.2 see [Str94, page 91, Example 5.13.1]. For the sentences 5.1.7-5.1.9 we refer to [Bog98, pages 42, 48, Sections 2.2-2.3, Theorem 2.3.1] and [DPZ92, pages 53-58, Section 2.3.2, Propositions 2.15, 2.18]. For Definition 5.1.10 and Proposition 5.1.12 see [Shi84, pages 299, 303, Sections II.13.4, II.13.6, Theorem 1].

5.1.2 Standard Brownian motion and the Wiener measure

Definition 5.1.13 A centered Gaussian system $\{W(t)\}_{t \geq 0}$ of random variables on a probability space $(\Omega, \mathcal{F}, \mathbb{P})$ is called *standard Brownian motion* if it has the following properties:

(i) For almost every $\omega \in \Omega$, $W(0)(\omega) = 0$;

(ii) If $n \in \mathbb{N}$ and $0 = t_0 < t_1 < \cdots < t_n$, then for $j = 1, \cdots, n$ the random variables $W(t_j) - W(t_{j-1})$ are independent;

(iii) For all $0 \leq s < t$, $W(t) - W(s)$ is a centered Gaussian random variable with variance $t - s$;

(iv) For almost every $\omega \in \Omega$ the sample path $t \mapsto W(t)(\omega)$ belongs to $C[0, \infty)$.

For a proof of the following theorem we refer to [KS91, page 70, Theorem 4.20].

Theorem 5.1.14 *Let $\Omega := C_0[0, \infty)$ be endowed with the metric d given by*

$$d(\omega_1, \omega_2) := \sum_{n=1}^{\infty} \frac{1}{2^n} \max_{t \in [0,n]} \min\big\{1, |\omega_1(t) - \omega_2(t)|\big\}, \quad \omega_1, \omega_2 \in C_0[0, \infty),$$

and let $\mathcal{F}$ be the Borel σ-algebra on (Ω, d). Let the collection $\{B(t)\}_{t\geq 0}$ of random variables on $(\Omega, \mathcal{F})$ be given by

$$B(t)(\omega) := \omega(t), \quad t \geq 0,\, \omega \in \Omega.$$

Then there exists a unique probability measure $\mathbb{P}$ on $(\Omega, \mathcal{F})$ such that $\{B(t)\}_{t\geq 0}$ is standard Brownian motion satisfying

$$\int_\Omega B(t)(\omega)\, B(s)(\omega)\, \mathbb{P}(\mathrm{d}\omega) = \min\{s,t\}, \quad s,t \geq 0.$$

Definition 5.1.15 The probability measure $\mathbb{P}$ in Theorem 5.1.14 is called the *Wiener measure.*

We can see standard Brownian motion $\{W(t)\}_{t\geq 0}$ on a probability space $(\Omega, \mathcal{F}, \mathbb{P})$ as a function $W : [0,\infty) \to L^2(\Omega)$. Even more, Definition 5.1.13(iii) implies the next corollary:

Corollary 5.1.16 *If $\{W(t)\}_{t\geq 0}$ is standard Brownian motion on a probability space $(\Omega, \mathcal{F}, \mathbb{P})$, then W belongs to $C_0^{0,\frac{1}{2}}([0,T]; L^2(\Omega))$ for every $T > 0$.*

Theorem 5.1.17 *Let $\{W(t)\}_{t\geq 0}$ be standard Brownian motion on a probability space $(\Omega, \mathcal{F}, \mathbb{P})$. Then for all $T > 0$ and $\gamma \in (0, \frac{1}{2})$ and for almost every $\omega \in \Omega$ the sample path $t \mapsto W(t)(\omega)$ belongs to $C_0^{0,\gamma}[0,T]$.*

Theorem 5.1.18 *Let $\{W(t)\}_{t\geq 0}$ be standard Brownian motion on a probability space $(\Omega, \mathcal{F}, \mathbb{P})$. Then for almost every $\omega \in \Omega$ the sample path $t \mapsto W(t)(\omega)$ is nowhere differentiable.*

For Theorems 5.1.17 and 5.1.18 see [KS91, pages 56, 110, Remark 2.12, Theorem 9.18].

5.1.3 The Wiener integral

Let $\Omega := C_0[0,\infty)$ be endowed with the metric d defined in Theorem 5.1.14 and let $\mathcal{F}$ be the Borel σ-algebra on (Ω, d). Let $\{B(t)\}_{t\geq 0}$ be standard Brownian motion on $(\Omega, \mathcal{F}, \mathbb{P})$ where $\mathbb{P}$ is the Wiener measure. Let $T > 0$.
If $f : [0,T] \to \mathbb{R}$ is a step function, that is, for some $n \in \mathbb{N}$, $c_1, \cdots, c_n \in \mathbb{R}$, and $0 = t_0 \leq t_1 \leq \cdots \leq t_n = T$, $f = \sum_{j=1}^n c_j \mathbf{1}_{[t_{j-1},t_j)}$, then the random variable $I(f) : \Omega \to \mathbb{R}$ is defined by

$$I(f) := \sum_{j=1}^n c_j \big(B(t_j) - B(t_{j-1})\big),$$

and has the property that $I(f)$ belongs to $L^2(\Omega)$ with $\|I(f)\|_{L^2(\Omega)} = \|f\|_{L^2(0,T)}$. Since the step functions are dense in $L^2(0,T)$, we have the following proposition, see [GN01, page 20, Theorem 2.5] for a proof.

Proposition 5.1.19 *The mapping $f \mapsto I(f)$ extends uniquely to an isometric isomorphism $I : L^2(0,T) \to L^2(\Omega)$.*

Definition 5.1.20 If f belongs to $L^2(0,T)$, then $I(f)$ is called the *Wiener integral of f with respect to $\{B(t)\}_{t\geq 0}$* and it is denoted by

$$\int_0^T f(t)\,\mathrm{d}B(t) := I(f).$$

Lemma 5.1.21 *If f belongs to $L^2(0,T)$, then for every $t \in [0,T]$, $\int_0^t f(s)\,\mathrm{d}B(s) = \int_0^T \mathbf{1}_{[0,t]}(s)f(s)\,\mathrm{d}B(s)$.*

Proposition 5.1.22 *Let f belong to $L^2(0,T)$.*

(i) *For every $t \in [0,T]$, $\int_0^t f(s)\,\mathrm{d}B(s)$ is a centered Gaussian variable with variance $\int_0^t (f(s))^2\,\mathrm{d}s$;*

(ii) *For every $g \in L^2(0,T)$,*

$$\mathbb{E}\left(\int_0^t f(s)\,\mathrm{d}B(s)\int_0^t g(s)\,\mathrm{d}B(s)\right) = \int_0^t f(s)g(s)\,\mathrm{d}s, \quad t \in [0,T];$$

(iii) *The collection $\left\{\int_0^t f(s)\,\mathrm{d}B(s)\right\}_{t\geq 0}$ is a centered Gaussian system and for all $s,t \in [0,T]$*

$$\mathbb{E}\left(\int_0^t f(\sigma)\,\mathrm{d}B(\sigma)\int_0^s f(\sigma)\,\mathrm{d}B(\sigma)\right) = \int_0^{\min\{s,t\}} (f(\sigma))^2\,\mathrm{d}\sigma.$$

For a proof of Proposition 5.1.22(i) see [GN01, page 20, Theorem 2.6].

5.1.4 The stochastic Fubini theorem

For a proof of the following results we refer to [GN01, pages 22-24, Lemma's 2.7-2.8, Theorem 2.9].

Lemma 5.1.23 *If f belongs to $L^2([0,T]\times[0,T])$, then there exists a Borel nullset $N \subseteq [0,T]$ such that for every $t \in [0,T]\backslash N$ the mapping $s \mapsto f(s,t)$ belongs to $L^2(0,T)$. If, in addition, $f_B : [0,T] \to L^2(\Omega)$ is given by*

$$f_B(t) := \begin{cases} \int_0^T f(s,t)\,\mathrm{d}B(s), & t \in [0,T]\backslash N, \\ 0, & t \in N, \end{cases}$$

then f_B belongs to $L^2(0,T;L^2(\Omega))$ with

$$\|f_B\|_{L^2(0,T;L^2(\Omega))} = \|f\|_{L^2([0,T]\times[0,T])}.$$

Even more, f_B belongs to $L^1(0,T;L^2(\Omega))$ with

$$\left\|\int_0^T f_B(t)\,\mathrm{d}t\right\|_{L^2(\Omega)} \leq \sqrt{T}\|f\|_{L^2([0,T]\times[0,T])}.$$

Lemma 5.1.24 *If f belongs to $L^2([0,T]\times[0,T])$ and if $F:[0,T]\to\mathbb{R}$ is given by*

$$F(s) := \int_0^T f(s,t)\,\mathrm{d}t, \quad s\in[0,T],$$

then F belongs to $L^2(0,T)$ with

$$\|F\|_{L^2(0,T)} \leq \sqrt{T}\|f\|_{L^2([0,T]\times[0,T])}.$$

Theorem 5.1.25 *If f belongs to $L^2([0,T]\times[0,T])$, then*

$$\int_0^T\left(\int_0^T f(s,t)\,\mathrm{d}B(s)\right)\mathrm{d}t = \int_0^T\left(\int_0^T f(s,t)\,\mathrm{d}t\right)\mathrm{d}B(s).$$

5.1.5 Weak solutions

Let $\Omega := C_0[0,\infty)$ be endowed with the metric d defined in Theorem 5.1.14 and let $\mathcal{F}$ be the Borel σ-algebra on (Ω,d). Let $\{B(t)\}_{t\geq 0}$ be standard Brownian motion on $(\Omega,\mathcal{F},\mathbb{P})$ where $\mathbb{P}$ is the Wiener measure, that is, for all $t\geq 0$ and $\omega\in\Omega$, $B(t)(\omega) := \omega(t)$, see Theorem 5.1.14.

Definition 5.1.26 The σ-algebra $\mathcal{P}_\infty$ on $[0,\infty)\times\Omega$ is the σ-algebra generated by subsets of the form $\{0\}\times\emptyset$, $\{0\}\times\Omega$, and $(s,t]\times F$, where $0\leq s<t<\infty$ and $F\in\sigma(B(s))$, the σ-algebra of subsets of Ω generated by $B(s)$.

Definition 5.1.27 A *predictable process* with values in a separable Hilbert space H is an H-valued stochastic process $\{X(t)\}_{t\geq 0}$ such that the mapping $(t,\omega)\mapsto X(t)(\omega)$ is strongly measurable from $([0,\infty)\times\Omega,\mathcal{P}_\infty)$ into $(H,\mathcal{B}(H))$.

Let $(H,\langle\cdot,\cdot\rangle,\|\cdot\|)$ be a separable Hilbert space. Let $\{T(t)\}_{t\geq 0}$ be an analytic semigroup with infinitesimal generator $A:\mathrm{D}(A)\subseteq H\to H$ and let $x\in H$.
We consider the following stochastic problem:

$$\begin{aligned} \tfrac{\mathrm{d}}{\mathrm{d}t}X(t) &= AX(t) + x\,\dot{B}(t), \quad t>0, \\ X(0) &= 0. \end{aligned} \tag{5.1}$$

To give problem (5.1) a meaning we interpret it as

$$X(t) = A\int_0^t X(s)\,\mathrm{d}s + xB(t), \quad t \geq 0.$$

Definition 5.1.28 A *weak solution* to problem (5.1) is an H-valued predictable process $\{X(t)\}_{t\geq 0}$ with the following properties:

(i) For every $T > 0$ we have that for almost every $\omega \in \Omega$ the mapping $t \mapsto X(t)(\omega)$ belongs to $L^1(0,T;H)$;

(ii) For all $y \in \mathrm{D}(A^*)$ and $t \geq 0$ we have for almost every $\omega \in \Omega$,

$$\big\langle X(t)(\omega), y\big\rangle = \int_0^t \big\langle X(s)(\omega), A^*y\big\rangle\,\mathrm{d}s + \omega(t)\langle x, y\rangle.$$

5.2 Weak and pathwise formulation of the problem

Let $\Omega := C_0[0,\infty)$ be endowed with the metric d defined in Theorem 5.1.14 and let $\mathcal{F}$ be the Borel σ-algebra on (Ω, d). Let $\{B(t)\}_{t\geq 0}$ be standard Brownian motion on $(\Omega, \mathcal{F}, \mathbb{P})$ where $\mathbb{P}$ is the Wiener measure, that is, for all $t \geq 0$ and $\omega \in \Omega$, $B(t)(\omega) := \omega(t)$, see Theorem 5.1.14.

We consider the stochastic linear Volterra integrodifferential equation

$$\begin{aligned} \frac{\mathrm{d}}{\mathrm{d}t}\int_{-\infty}^t a(t-s)U(s)\,\mathrm{d}s &= f(t) + \sigma\dot{B}(t), && t > 0,\\ U(t) &= u_0(t), && t \leq 0, \end{aligned} \tag{5.2}$$

where $\sigma \in \mathbb{R}$, where the kernel $a : (0,\infty) \to \mathbb{R}$ belongs to the class $\mathcal{K}$, where the function $u_0 : (-\infty, 0] \to \mathbb{R}$ satisfies Hypothesis 3.6.2, and where the function $f : [0,\infty) \to \mathbb{R}$ belongs to $L^1(0,T)$ for every $T > 0$. To give problem (5.2) a meaning we interpret it as the stochastic first kind linear $L^2(\Omega)$-valued Volterra equation

$$\int_0^t a(t-s)U(s)\,\mathrm{d}s = h(t)\mathbf{1}_\Omega + \sigma B(t), \quad t \geq 0, \tag{5.3}$$

where the function $h : [0,\infty) \to \mathbb{R}$ is defined by

$$h(t) := \int_{-\infty}^0 a(-s)u_0(s)\,\mathrm{d}s - \int_{-\infty}^0 a(t-s)u_0(s)\,\mathrm{d}s + \int_0^t f(s)\,\mathrm{d}s, \quad t \geq 0.$$

Note that h is well-defined by the assumptions on u_0 and f and that h belongs to $C_0[0,\infty)$.

Definition 5.2.1 A *solution* to problem (5.2) is a strongly Borel measurable function $U : \mathbb{R} \to L^2(\Omega)$ with the following properties:

(i) $\int_0^t a(t-s)\|U(s)\|_{L^2(\Omega)}\,\mathrm{d}s < \infty$ for every $t \geq 0$;

(ii) U satisfies (5.3) for every $t \geq 0$;

(iii) $U(t) = u_0(t)\mathbf{1}_\Omega$ for almost every $t \leq 0$.

Analogous to Section 3.1 we have that a solution U to problem (5.2) belongs to $L^1\big(0,T;L^2(\Omega)\big)$ for every $T > 0$. Moreover, we observe that if U is a solution to problem (5.2), then $U|_{[0,\infty)}$ is a solution to problem (5.3) in the sense of Definition 4.3.1. Therefore Theorem 4.3.2 states the following:

Theorem 5.2.2 *Problem (5.2) admits at most one solution.*

Note that when $\sigma = 0$ the following proposition holds and hence, the results of Chapter 3 apply.

Proposition 5.2.3 *If the scalar linear Volterra integrodifferential equation*

$$\begin{aligned} \frac{\mathrm{d}}{\mathrm{d}t}\int_{-\infty}^t a(t-s)u(s)\,\mathrm{d}s &= f(t), \quad t>0, \\ u(t) &= u_0(t), \quad t \leq 0, \end{aligned} \tag{5.4}$$

admits a solution $u : \mathbb{R} \to \mathbb{R}$ in the sense of Definition 3.1.1, then problem (5.2) with $\sigma = 0$ admits a unique solution U_{det}, and U_{det} has the representation

$$U_{\mathrm{det}}(t) = u(t)\mathbf{1}_\Omega, \quad t \in \mathbb{R}. \tag{5.5}$$

Now we consider problem (5.2) with $\sigma = 1$ and f and u_0 identically zero, that is,

$$\begin{aligned} \frac{\mathrm{d}}{\mathrm{d}t}\int_0^t a(t-s)U(s)\,\mathrm{d}s &= \dot{B}(t), \quad t>0, \\ U(t) &= 0, \qquad t \leq 0, \end{aligned} \tag{5.6}$$

which is interpreted as

$$\int_0^t a(t-s)U(s)\,\mathrm{d}s = B(t), \quad t \geq 0.$$

Theorem 5.2.4 *If a belongs to $\mathcal{K}$ with $\alpha(a) \in (\frac{1}{2},1)$, then problem (5.6) admits a unique solution U_B. Moreover, U_B belongs to $C^{0,\zeta}\big([0,T];L^2(\Omega)\big)$ for all $T>0$ and $\zeta \in \big(0,\alpha(a)-\frac{1}{2}\big)$, and U_B has the representation*

$$U_B(t) = \begin{cases} \frac{\mathrm{d}}{\mathrm{d}t}\int_0^t J\big((\beta I_H - A)S(t-s)\pi\big)B(s)\,\mathrm{d}s, & t \geq 0, \\ 0, & t < 0. \end{cases} \tag{5.7}$$

PROOF: We use Corollary 5.1.16 and apply Theorem 4.3.3 with $X := L^2(\Omega)$, $h := B$, and $\gamma := \frac{1}{2}$. □

Corollary 5.2.5 *If a belongs to $\mathcal{K}$ with $\alpha(a) \in (\frac{1}{2}, 1)$ and if problem (5.4) admits a solution $u : \mathbb{R} \to \mathbb{R}$ in the sense of Definition 3.1.1, then problem (5.2) admits a unique solution U, and U has the representation*

$$U(t) = U_{\text{det}}(t) + \sigma U_B(t), \quad t \in \mathbb{R},$$

where U_{det} and U_B are given by (5.5) respectively (5.7).

5.3 Stochastic convolution

Let $\Omega := C_0[0,\infty)$ be endowed with the metric d defined in Theorem 5.1.14 and let $\mathcal{F}$ be the Borel σ-algebra on (Ω, d). Let $\{B(t)\}_{t\geq 0}$ be standard Brownian motion on $(\Omega, \mathcal{F}, \mathbb{P})$ where $\mathbb{P}$ is the Wiener measure, that is, for all $t \geq 0$ and $\omega \in \Omega$, $B(t)(\omega) := \omega(t)$, see Theorem 5.1.14.

5.3.1 The direct approach

Let $a : (0,\infty) \to \mathbb{R}$ belong to the class $\mathcal{K}$ and let $b : (0,\infty) \to \mathbb{R}$ denote the resolvent of the first kind of a. If $\alpha(a) \in (\frac{1}{2}, 1)$, then b belongs to $L^2(0,T)$ for every $T > 0$ by Theorem 3.7.7. In that case we define the function $Z : [0,\infty) \to L^2(\Omega)$ by

$$Z(t) := \begin{cases} \int_0^t b(t-s)\,\mathrm{d}B(s), & t > 0, \\ 0, & t = 0. \end{cases} \tag{5.8}$$

Note that Z is well-defined as a Wiener integral. Moreover, Proposition 5.1.22(iii) implies that $\{Z(t)\}_{t\geq 0}$ is a centered Gaussian system with covariance

$$\mathbb{E}\big(Z(t)Z(s)\big) = \int_0^{\min\{s,t\}} \big(b(\sigma)\big)^2\,\mathrm{d}\sigma, \quad s,t \geq 0.$$

Lemma 5.3.1 *If a belongs to $\mathcal{K}$ with $\alpha(a) \in (\frac{1}{2}, 1)$, then Z, given by (5.8), is continuous.*

PROOF: The continuity of Z at $t = 0$ follows from

$$\lim_{t\downarrow 0} \left\| Z(t) \right\|_{L^2(\Omega)}^2 = \lim_{t\downarrow 0} \int_0^t \big(b(s)\big)^2\,\mathrm{d}s = 0.$$

Now let $t > 0$ and $T > t$. Let $\{t_n\}_{n=1}^{\infty}$ be a sequence in $[0,T]$ such that $\lim_{n\to\infty} t_n = t$. Then we have for $n = 1, 2, \cdots$,

$$\begin{aligned}
&\left\|Z(t_n) - Z(t)\right\|^2_{L^2(\Omega)} \\
&= \left\| \int_0^T \left(\mathbf{1}_{[0,t_n)}(s)b(t_n - s) - \mathbf{1}_{[0,t)}(s)b(t-s)\right) \mathrm{d}B(s) \right\|^2_{L^2(\Omega)} \\
&= \int_0^T \left(\mathbf{1}_{[0,t_n)}(s)b(t_n - s) - \mathbf{1}_{[0,t)}(s)b(t-s)\right)^2 \mathrm{d}s \\
&= \int_0^{t_n} \left(b(\sigma)\right)^2 \mathrm{d}\sigma + \int_0^t \left(b(\sigma)\right)^2 \mathrm{d}\sigma - 2\int_0^{\min\{t_n,t\}} b(\sigma)\, b\big(\sigma + |t - t_n|\big)\, \mathrm{d}\sigma \\
&= 2\int_0^{\min\{t_n,t\}} b(\sigma)\, \big(b(\sigma) - b\big(\sigma + |t-t_n|\big)\big)\, \mathrm{d}\sigma + \int_{\min\{t_n,t\}}^{\max\{t_n,t\}} \left(b(\sigma)\right)^2 \mathrm{d}\sigma.
\end{aligned}$$

Therefore the continuity of Z at t is a result of Lebesgue's dominated convergence theorem. □

Proposition 5.3.2 *If a belongs to $\mathcal{K}$ with $\alpha(a) \in (\frac{1}{2}, 1)$ and Z is given by (5.8), then*

$$\int_0^t a(t-s)Z(s)\,\mathrm{d}s = B(t), \quad t \geq 0.$$

PROOF: For $n = 1, 2, \cdots$ let $a_n : [0,\infty) \to \mathbb{R}$ be defined by

$$a_n(t) := a(t + \tfrac{1}{n}), \quad t \geq 0.$$

Then $\lim_{n\to\infty} \|a_n - a\|_{L^1(0,T)} = 0$ and a_n belongs to $L^2(0,T)$ for $n = 1, 2, \cdots$ and every $T > 0$. Let $t > 0$. For $n = 1, 2, \cdots$ we define $f_n : [0,t] \times [0,t] \to \mathbb{R}$ by

$$f_n(s,\sigma) := \begin{cases} a_n(t-s)b(s-\sigma), & 0 \leq \sigma < s < t, \\ 0, & 0 \leq s \leq \sigma \leq t. \end{cases}$$

Note that f_n is Borel measurable and belongs to $L^2([0,t] \times [0,t])$ for $n = 1, 2, \cdots$. Thus we can apply Theorem 5.1.25 to obtain

$$\begin{aligned}
&\int_0^t a_n(t-s)Z(s)\,\mathrm{d}s = \int_{s=0}^t a_n(t-s)\left(\int_{\sigma=0}^s b(s-\sigma)\,\mathrm{d}B(\sigma)\right)\mathrm{d}s \\
&= \int_{s=0}^t \left(\int_{\sigma=0}^t f_n(s,\sigma)\,\mathrm{d}B(\sigma)\right)\mathrm{d}s = \int_{\sigma=0}^t \left(\int_{s=0}^t f_n(s,\sigma)\,\mathrm{d}s\right)\mathrm{d}B(\sigma) \\
&= \int_{\sigma=0}^t \left(\int_{s=\sigma}^t a_n(t-s)b(s-\sigma)\,\mathrm{d}s\right)\mathrm{d}B(\sigma) \\
&= \int_{\sigma=0}^t \left(\int_{s=0}^{t-\sigma} a_n(t-\sigma-s)b(s)\,\mathrm{d}s\right)\mathrm{d}B(\sigma), \qquad n = 1, 2, \cdots.
\end{aligned}$$

To finish the proof it is sufficient to show that we have the following limits in $L^2(\Omega)$:

$$\lim_{n\to\infty} \int_0^t a_n(t-s)Z(s)\,\mathrm{d}s = \int_0^t a(t-s)Z(s)\,\mathrm{d}s, \tag{5.9}$$

$$\lim_{n\to\infty} \int_{\sigma=0}^t \left(\int_{s=0}^{t-\sigma} a_n(t-\sigma-s)b(s)\,\mathrm{d}s \right) \mathrm{d}B(\sigma) = B(t). \tag{5.10}$$

Since Z belongs to $C\big([0,t];L^2(\Omega)\big)$ by Lemma 5.3.1 and $\lim_{n\to\infty} \|a_n - a\|_{L^1(0,t)} = 0$, the theory of convolutions implies that (5.9) holds. Moreover, as b belongs to $L^2(0,t)$ we have the next limit in $L^2(0,t)$:

$$\lim_{n\to\infty} \int_0^{t-\sigma} \big(a_n(t-\sigma-s) - a(t-\sigma-s)\big)b(s)\,\mathrm{d}s = 1, \quad \sigma \in [0,t). \tag{5.11}$$

Since $B(t) = \int_0^t \mathrm{d}B(\sigma)$ and $\int_0^{t-\sigma} a(t-\sigma-s)b(s)\,\mathrm{d}s = 1$ for every $\sigma \in [0,t)$, we have

$$\begin{aligned}
&\left\| \int_{\sigma=0}^t \left(\int_{s=0}^{t-\sigma} a_n(t-\sigma-s)b(s)\,\mathrm{d}s \right) \mathrm{d}B(\sigma) - B(t) \right\|_{L^2(\Omega)} \\
&= \left\| \int_{\sigma=0}^t \left(\int_{s=0}^{t-\sigma} \big(a_n(t-\sigma-s) - a(t-\sigma-s)\big)b(s)\,\mathrm{d}s \right) \mathrm{d}B(\sigma) \right\|_{L^2(\Omega)} \\
&= \left\| \int_{s=0}^{t-\sigma} \big(a_n(t-\sigma-s) - a(t-\sigma-s)\big)b(s)\,\mathrm{d}s \right\|_{L^2(0,t)}.
\end{aligned}$$

Now (5.10) follows from (5.11). This finishes the proof. □

Lemma 5.3.3 *If a belongs to $\mathcal{K}$ with $\alpha(a) \in (\frac{1}{2}, 1)$, if U_B is the solution to problem (5.6), and if Z is given by (5.8), then*

$$U_B(t) = Z(t), \quad t \geq 0.$$

PROOF: First we observe that U_B and Z are continuous on $[0,\infty)$ by Theorem 5.2.4 respectively Lemma 5.3.1. Moreover, by Definition 5.2.1 and Proposition 5.3.2 we have

$$\int_0^t a(t-s)U_B(s)\,\mathrm{d}s = B(t) = \int_0^t a(t-s)Z(s)\,\mathrm{d}s, \quad t \geq 0.$$

This implies that

$$\int_0^t a(t-s)\big\langle U_B(s) - Z(s), x\big\rangle_{L^2(\Omega)}\,\mathrm{d}s = 0, \quad t \geq 0,\, x \in L^2(\Omega).$$

Now it follows from Corollory 1.2.2 that $\langle U_B(s) - Z(s), x \rangle_{L^2(\Omega)} = 0$ for every $x \in L^2(\Omega)$ and almost every $t > 0$. Hence, using the continuity of U_B and Z on $[0, \infty)$ we have $U_B(t) = Z(t)$ for every $t > 0$. □

We remark that due to Theorem 3.7.7 and the representation (5.7) of the solution U_B to problem (5.6), Lemma 5.3.3 states that if a belongs to $\mathcal{K}$ with $\alpha(a) \in (\frac{1}{2}, 1)$, then

$$\frac{\mathrm{d}}{\mathrm{d}t} \int_0^t b(t-s)B(s)\,\mathrm{d}s = \int_0^t b(t-s)\,\mathrm{d}B(s), \quad t \geq 0.$$

Theorem 5.3.4 *Let a belong to $\mathcal{K}$ with $\alpha(a) \in (\frac{1}{2}, 1)$ and let U_B be the solution to problem (5.6). Let u_0 satisfy Hypothesis 3.6.2, let $f : [0, \infty) \to \mathbb{R}$ be such that problem (5.2) with $\sigma = 0$ admits a solution U_{det}, and let $\sigma \in \mathbb{R}$. Then $U : \mathbb{R} \to L^2(\Omega)$, given by*

$$U(t) := U_{\mathrm{det}}(t) + \sigma U_B(t), \quad t \in \mathbb{R},$$

is the unique solution to problem (5.2). Moreover, $\{U(t)\}_{t \geq 0}$ is a Gaussian system with mean

$$\mathbb{E}U(t) = \int_0^t \big(b(s)\big)^2\,\mathrm{d}s, \quad t \geq 0,$$

and covariance

$$\mathbb{E}\Big(\big(U(t) - \mathbb{E}U(t)\big)\big(U(s) - \mathbb{E}U(s)\big)\Big) = \int_0^{\min\{s,t\}} \big(b(\sigma)\big)^2\,\mathrm{d}\sigma, \quad t, s \geq 0.$$

PROOF: The theorem follows from Proposition 5.2.3, Theorem 5.2.4, and Lemma 5.3.3. □

5.3.2 The semigroup approach

In this section we follow our semigroup approach to study problem (5.6). Analogous to Section 3.7 we first consider the abstract stochastic problem

$$\begin{aligned} &\tfrac{\mathrm{d}}{\mathrm{d}t}\Psi(t) = A_{-1}\Psi(t) + (\beta I_H - A_{-1})\pi\,\dot{B}(t), \quad t > 0, \\ &\Psi(0) = 0. \end{aligned} \tag{5.12}$$

We recall that the function $\pi \in H$ is defined by

$$\pi(\kappa) := \frac{1}{\beta\hat{a}(\beta)}\frac{1}{\kappa + \beta}, \quad \kappa \geq 0.$$

Moreover, Proposition 3.5.7 implies that π belongs to $(H, \mathrm{D}(A))_{\theta,2}$ for every $\theta \in (0, 1)$ such that $\theta < \frac{1+\alpha(a)}{2}$ and hence, $(\beta I_H - A_{-1})\pi$ belongs to $(H_{-1}, H)_{\theta,2}$ by Proposition 1.4.18. Therefore we can consider problem (5.12) in the space $(H_{-1}, H)_{\theta,2}$.

Proposition 5.3.5 *Problem (5.12) admits a unique $(H_{-1}, H)_{\theta,2}$-valued weak solution $\{\Psi(t)\}_{t\geq 0}$ for every $\theta \in (0,1)$ such that $\theta < \frac{1+\alpha(a)}{2}$. Moreover, the weak solution is a centered Gaussian process with the representation*

$$\Psi(t) = \int_0^t (\beta I_H - A)S(t-s)\pi \, \mathrm{d}B(s), \quad t \geq 0.$$

PROOF: We refer to [DPZ92, pages 121, 119, Theorems 5.4, 5.2], noting that for all $T > 0$ and $\theta \in (0,1)$ such that $\theta < \frac{1+\alpha(a)}{2}$,

$$\begin{aligned} &\int_0^T \mathrm{Tr}\left(S_{-1}(t)(\beta I_H - A_{-1})\pi \langle S_{-1}^*(t)\,\cdot, (\beta I_H - A_{-1})\pi\rangle_{(H_{-1},H)_{\theta,2}}\right) \mathrm{d}t \\ &= \int_0^T \big\|S_{-1}(t)(\beta I_H - A_{-1})\pi\big\|^2_{(H_{-1},H)_{\theta,2}} \mathrm{d}t \\ &\leq T\big\|(\beta I_H - A_{-1})\pi\big\|^2_{(H_{-1},H)_{\theta,2}} \max\left\{\|S_{-1}(t)\|^2_{\mathcal{L}((H_{-1},H)_{\theta,2})};\, t \in [0,T]\right\} < \infty. \end{aligned}$$

□

For a proof of the following proposition see [DPZ92, pages 133, 134, Propositions 5.15, 5.16].

Proposition 5.3.6 *If a belongs to $\mathcal{K}$ with $\alpha(a) \in (0,1)$, then for all $\theta, \eta \in (0,1)$ such that $\eta < \theta < \frac{1+\alpha(a)}{2}$ and $\theta - \eta > \frac{1}{2}$ the weak solution $\{\Psi(t)\}_{t\geq 0}$ to problem (5.12) is an $(H, \mathrm{D}(A))_{\eta,2}$-valued centered Gaussian process. Moreover, for all $T > 0$ and $\zeta \in \left(0, \theta - \eta - \frac{1}{2}\right)$ we have that for almost every $\omega \in \Omega$ the mapping $t \mapsto \Psi(t)(\omega)$ belongs to $C^\zeta\big([0,T]; (H, \mathrm{D}(A))_{\eta,2}\big)$.*

Now we return to problem (5.6).

Theorem 5.3.7 *If a belongs to $\mathcal{K}$ with $\alpha(a) \in (\frac{1}{2}, 1)$, then problem (5.6) admits a unique solution U and for all $T > 0$ and $\zeta \in \left(0, \alpha(a) - \frac{1}{2}\right)$ we have that for almost every $\omega \in \Omega$ the mapping $t \mapsto U(t)(\omega)$ belongs to $C^{0,\zeta}[0,T]$. Moreover, if $\theta, \eta \in (0,1)$ are such that $\frac{1-\alpha(a)}{2} < \eta < \theta < \frac{1+\alpha(a)}{2}$ and $\theta - \eta > \frac{1}{2}$, and if $\{\Psi(t)\}_{t\geq 0}$ is the $(H, \mathrm{D}(A))_{\eta,2}$-valued weak solution to problem (5.12), then U has the representation*

$$U(t) = \begin{cases} J(\Psi(t)), & t \geq 0, \\ 0, & t < 0. \end{cases}$$

The proof of this theorem is an application of the Hilbert-valued theory given in Section 4.1 to the special case $X := L^2(\Omega)$.

With this result we establish a state-space setting for stochastic convolution equations, so that the solutions can be considered within the framework of Markov processes. This opens the door for investigations in stationary asymptotic distributions, transition probabilities, and the Kolmogorov equation.

Bibliography

[Ama95] H. Amann. *Linear and quasilinear parabolic problems.* Birkhäuser Verlag, Basel, 1995.

[Baj01] E.G. Bajlekova. *Fractional evolution equations in Banach spaces.* Ph.D. thesis, Technische Universiteit Eindhoven, Eindhoven, 2001.

[BL76] J. Bergh and J. Löfström. *Interpolation spaces. An introduction.* Springer-Verlag, Berlin, 1976.

[Bog98] V.I. Bogachev. *Gaussian measures.* American Mathematical Society, Providence, RI, 1998.

[Bon01] S. Bonaccorsi. Onsager-Machlup functional for Volterra equations perturbed by noise. *Università degli Studi di Trento*, UTM 602, 2001.

[BT01] S. Bonaccorsi and L. Tubaro. Mittag-Leffler's function and stochastic linear Volterra equations of convolutions type. *Università degli Studi di Trento*, UTM 604, 2001.

[BvN00] Z. Brzeźniak and J. van Neerven. Stochastic convolution in separable Banach spaces and the stochastic linear Cauchy problem. *Studia Math.*, 143:43–74, 2000.

[CDH] Ph. Clément, W. Desch, and K.W. Homan. An analytic semigroup setting for a class of Volterra equations. To appear in *Journal of Integral Equations & Applications.*

[CDH02] Ph. Clément, W. Desch, and K.W. Homan. An analytic semigroup setting for a class of Volterra equations. *Optimierung und Kontrolle, Karl-Franzens-Universität und Technische Universität Graz,* SFB F003 Bericht Nr. 242, 2002.

[CDP97] Ph. Clément and G. Da Prato. White noise perturbation of the heat equation in materials with memory. *Dynam. Systems Appl.*, 6:441–460, 1997.

[CDPP97] Ph. Clément, G. Da Prato, and J. Prüss. White noise perturbation of the equations of linear parabolic viscoelasticity. *Rend. Istit. Mat. Univ. Trieste*, 29:207–220, 1997.

[CHA+87] Ph. Clément, H.J.A.M. Heijmans, S. Angenent, C.J. van Duijn, and B. de Pagter. *One-parameter semigroups.* North-Holland Publishing Co., Amsterdam, 1987.

[CN81] Ph. Clément and J.A. Nohel. Asymptotic behavior of solutions of nonlinear Volterra equations with completely positive kernels. *SIAM J. Math. Anal.*, 12:514–535, 1981.

[CP90] Ph. Clément and J. Prüss. Completely positive measures and Feller semigroups. *Math. Ann.*, 287:73–105, 1990.

[DG89] W. Desch and R. Grimmer. Singular relaxation moduli and smoothing in three-dimensional viscoelasticity. *Trans. Amer. Math. Soc.*, 314:381–404, 1989.

[DH] W. Desch and K.W. Homan. Pointwise versions of solutions to Cauchy problems in L^p-spaces. To appear in *Rendiconti dell'Istituto di Matematica dell'Università di Trieste.*

[DH02] W. Desch and K.W. Homan. Pointwise versions of solutions to Cauchy problems in L^p-spaces. *Optimierung und Kontrolle, Karl-Franzens-Universität und Technische Universität Graz*, SFB F003 Bericht Nr. 250, 2002.

[DM88] W. Desch and R.K. Miller. Exponential stabilization of Volterra integral equations with singular kernels. *J. Integral Equations Appl.*, 1:397–433, 1988.

[DPI85] G. Da Prato and M. Iannelli. Existence and regularity for a class of integro-differential equations of parabolic type. *J. Math. Anal. Appl.*, 112:36–55, 1985.

[DPZ92] G. Da Prato and J. Zabczyk. *Stochastic equations in infinite dimensions.* Cambridge University Press, Cambridge, 1992.

[DPZ96] G. Da Prato and J. Zabczyk. *Ergodicity for infinite dimensional systems.* Cambridge University Press, Cambridge, 1996.

[DPZ02] G. Da Prato and J. Zabczyk. *Second order partial differential equations in Hilbert spaces.* Cambridge University Press, Cambridge, 2002.

[DU77] J. Diestel and J.J. Uhl, Jr. *Vector measures.* American Mathematical Society, Providence, R.I., 1977.

[EG77] R.E. Edwards and G.I. Gaudry. *Littlewood-Paley and multiplier theory.* Springer-Verlag, Berlin, 1977.

[Egb92] P. Egberts. *On the sum of accretive operators.* Ph.D. thesis, Technische Universiteit Delft, Delft, 1992.

[Eng96] H. Engler. Global smooth solutions for a class of parabolic integro-differential equations. *Trans. Amer. Math. Soc.*, 348:267–290, 1996.

[GL95] G. Gripenberg and S.-O. Londen. Fractional derivatives and smoothing in nonlinear conservation laws. *Differential Integral Equations*, 8:1961–1976, 1995.

[GLS90] G. Gripenberg, S.-O. Londen, and O.J. Staffans. *Volterra integral and functional equations.* Cambridge University Press, Cambridge, 1990.

[GN01] O.W. van Gaans and J.M.A.M. van Neerven. Stochastic differential equations. *Mathematical Institute University of Leiden*, Report Nr. MI 2001-18, 2001.

[Gri80] G. Gripenberg. On Volterra equations of the first kind. *Integral Equations Operator Theory*, 3:473–488, 1980.

[Gri85] G. Gripenberg. Volterra integro-differential equations with accretive nonlinearity. *J. Differential Equations*, 60(1):57–79, 1985.

[GV91] R. Gorenflo and S. Vessella. *Abel integral equations.* Springer-Verlag, Berlin, 1991.

[HL28] G.H. Hardy and J.E. Littlewood. Some properties of fractional integrals. *Math. Zeitschrift*, 27:565–606, 1928.

[Jak01] V.G. Jakubowski. *Nonlinear elliptic-parabolic integrodifferential equations with L^1-data: existence, uniqueness, asymptotics.* Ph.D. thesis, Universität Essen, Essen, 2001.

[Kre78] E. Kreyszig. *Introductory functional analysis with applications.* John Wiley & Sons, New York-London-Sydney, 1978.

[KS91] I. Karatzas and S.E. Shreve. *Brownian motion and stochastic calculus.* Springer-Verlag, New York, second edition, 1991.

[KZ00a] A. Karczewska and J. Zabczyk. Regularity of solutions to stochastic Volterra equations. *Atti Accad. Naz. Lincei Cl. Sci. Fis. Mat. Natur. Rend. Lincei (9) Mat. Appl.*, 11:141–154, 2000.

[KZ00b] A. Karczewska and J. Zabczyk. Stochastic PDE's with function-valued solutions. In *Infinite dimensional stochastic analysis (Amsterdam, 1999)*, pages 197–216. R. Neth. Acad. Arts Sci., Amsterdam, 2000.

[LN84] S.-O. Londen and J.A. Nohel. Nonlinear Volterra integro-differential equation occurring in heat flow. *J. Integral Equations*, 6:11–50, 1984.

[Lun90] A. Lunardi. On the linear heat equation with fading memory. *SIAM J. Math. Anal.*, 21:1213–1224, 1990.

[Lun91] A. Lunardi. *Interpolation theory.* Scuola Normale Superiore, Pisa, 1991.

[Lun95] A. Lunardi. *Analytic semigroups and optimal regularity in parabolic problems.* Birkhäuser Verlag, Basel, 1995.

[MCSA01] C. Martínez Carracedo and M. Sanz Alix. *The theory of fractional powers of operators.* North-Holland Publishing Co., Amsterdam, 2001.

[Mil74] R.K. Miller. Linear Volterra integrodifferential equations as semigroups. *Funkcial. Ekvac.*, 17:39–55, 1974.

[Noh81] J.A. Nohel. Nonlinear Volterra equations for heat flow in materials with memory. In *Integral and functional differential equations (Proc. Conf., West Virginia Univ., Morgantown, W. Va., 1979)*, pages 3–82. Dekker, New York, 1981.

[Nun71] J.W. Nunziato. On heat conduction in materials with memory. *Quart. Appl. Math.*, 29:187–204, 1971.

[Paz83] A. Pazy. *Semigroups of linear operators and applications to partial differential equations.* Springer-Verlag, New York, second edition, 1983.

[Prü93] J. Prüss. *Evolutionary integral equations and applications.* Birkhäuser Verlag, Basel, 1993.

[RHN87] M. Renardy, W.J. Hrusa, and J.A. Nohel. *Mathematical problems in viscoelasticity.* Longman Scientific & Technical, Harlow, 1987.

[Roy88] H.L. Royden. *Real analysis.* Macmillan Publishing Company, New York, third edition, 1988.

[Shi84] A.N. Shiryayev. *Probability.* Springer-Verlag, New York, 1984.

[Sta94] O.J. Staffans. Well-posedness and stabilizability of a viscoelastic equation in energy space. *Trans. Amer. Math. Soc.*, 345:527–575, 1994.

[Ste70] E.M. Stein. *Topics in harmonic analysis related to the Littlewood-Paley theory.* Princeton University Press, Princeton, N.J., 1970.

[Str94] K.R. Stromberg. *Probability for analysts.* Chapman & Hall Probability Series. Chapman & Hall, New York, 1994.

[Tay65] A.E. Taylor. *General theory of functions and integration.* Blaisdell Publishing Co. Ginn and Co. , New York, 1965.

[Tri78] H. Triebel. *Interpolation theory, function spaces, differential operators.* North-Holland Publishing Co., Amsterdam, 1978.

[Wid41] D.V. Widder. *The Laplace transform.* Princeton University Press, Princeton, N. J., 1941.

[WW52] E.T. Whittaker and G.N. Watson. *A course of modern analysis.* Cambridge University Press, New York, fourth edition, 1952.

[WZ77] R.L. Wheeden and A. Zygmund. *Measure and integral; an introduction to real analysis.* Marcel Dekker Inc., New York, 1977.

[Yos95] K. Yosida. *Functional analysis.* Springer-Verlag, Berlin, sixth edition, 1995.

[Zac03] R. Zacher. *Quasilinear parabolic problems with nonlinear boundary conditions.* Ph.D. thesis, Martin-Luther-Universität Halle-Wittenberg, Halle, 2003.

Summary

of the thesis:

An analytic semigroup approach to convolution Volterra equations

This thesis is concerned with Volterra integrodifferential equations of convolution type with completely monotonic kernels. A motivation for studying this type of equations emanates from models involving memory, for example when considering heat flow in a material with memory. The main objective is to provide an analytic semigroup setting for these equations, based on the complete monotonicity of the kernel. Bernstein's theorem allows for rewriting the Volterra equation into an abstract Cauchy problem in an appropriate Hilbert space, in such a way that the operator governing this problem generates an analytic semigroup. Then the solution to the abstract Cauchy problem, together with interpolation methods, is used to obtain existence and regularity of solutions to the Volterra equation, as well as a representation formula.

The core of the thesis is Chapter 3, where we consider the scalar linear Volterra integrodifferential equation

$$\frac{\mathrm{d}}{\mathrm{d}t}\int_{-\infty}^{t} a(t-s)u(s)\,\mathrm{d}s = f(t), \quad t>0,$$
$$u(t) = u_0(t), \quad t \le 0.$$

When f is identically zero, this equation can be rewritten into the homogeneous abstract Cauchy problem

$$\begin{aligned} \tfrac{\mathrm{d}}{\mathrm{d}t}\varphi(t) &= A\varphi(t), \quad t>0, \\ \varphi(0) &= \psi_0, \end{aligned}$$

where A is the infinitesimal generator of an analytic semigroup in a Hilbert space H. By means of a linear functional acting on the solution φ to the homogeneous abstract Cauchy problem we obtain a solution u to the Volterra equation.

When f is not identically zero, it turns out that we cannot use the homogeneous abstract Cauchy problem as a semigroup setting. Instead, we think of an inhomogeneous abstract Cauchy problem in a larger space than H. We then solve this problem in an appropriate interpolation space $(H, \mathrm{D}(A))_{\eta,2}$ with $\eta \in (0,1)$, and use its solution to obtain a solution to the Volterra equation.
One aspect of our semigroup approach is not immediately clear. A solution φ to the homogeneous abstract Cauchy problem is an equivalence class in the Hilbert space, but we shall need a version of the equivalence class to show that the constructed function u is a solution to the Volterra equation. This problem is considered in Chapter 2.
In Chapter 4 we study three different types of Volterra integrodifferential equations in a separable Hilbert space X. Section 4.1 treats the X-valued equivalent of the scalar linear Volterra equation. In Section 4.2 we discuss the semilinear X-valued Volterra equation

$$\begin{aligned} \frac{\mathrm{d}}{\mathrm{d}t}\int_{-\infty}^{t} a(t-s)u(s)\,\mathrm{d}s &= f(t,u(t)), \quad t>0, \\ u(t) &= u_0(t), \qquad t \le 0. \end{aligned}$$

Section 4.3 is concerned with the first kind linear X-valued Volterra equation

$$\int_0^t a(t-s)u(s)\,\mathrm{d}s = h(t), \quad t \ge 0.$$

We are motivated to study this equation by the application of a stochastic perturbation of the linear Volterra equation in Chapter 5. In this chapter we consider the stochastic linear Volterra integrodifferential equation

$$\begin{aligned} \frac{\mathrm{d}}{\mathrm{d}t}\int_{-\infty}^{t} a(t-s)U(s)\,\mathrm{d}s &= f(t) + \sigma\dot{B}(t), \quad t>0, \\ U(t) &= u_0(t), \qquad t \le 0, \end{aligned}$$

where $\{B(t)\}_{t\ge0}$ is standard Brownian motion.

K.W. Homan,
Delft, February 2003

Samenvatting

van het proefschrift:

Een analytische halfgroep benadering voor convolutie Volterra vergelijkingen

Dit proefschrift gaat over Volterra integrodifferentiaalvergelijkingen van convolutie type met volledig monotone kernen. Een beweegreden om dit type vergelijkingen te bestuderen komt voort uit modellen die het verleden bevatten, bijvoorbeeld een model dat de warmtestroming beschrijft in een materiaal met geheugen. Het streven is om een analytische halfgroep benadering voor deze vergelijkingen op te zetten, uitgaande van de volledig monotoniciteit van de kern. Middels de stelling van Bernstein kan de Volterra vergelijking herschreven worden tot een abstract Cauchy probleem in een geschikte Hilbert ruimte, zodanig dat de operator die bepalend is voor het probleem een analytische halfgroep genereert. Vervolgens wordt de oplossing voor het abstracte Cauchy probleem, in combinatie met interpolatiemethoden, gebruikt om zowel existentie en regulariteit van oplossingen voor de Volterra vergelijking te verkrijgen, alsmede een representatie.
De kern van het proefschrift is Hoofdstuk 3, waarin we de volgende scalaire lineaire Volterra integrodifferentiaalvergelijking beschouwen:

$$\frac{\mathrm{d}}{\mathrm{d}t}\int_{-\infty}^{t} a(t-s)u(s)\,\mathrm{d}s = f(t), \quad t>0,$$
$$u(t) = u_0(t), \quad t \le 0.$$

In geval f identiek nul is, kan deze vergelijking herschreven worden tot het homogenene abstracte Cauchy probleem

$$\begin{aligned} \tfrac{\mathrm{d}}{\mathrm{d}t}\varphi(t) &= A\varphi(t), \quad t>0,\\ \varphi(0) &= \psi_0, \end{aligned}$$

waar A de infinitesimale generator van een analytische halfgroep in een Hilbert ruimte H is. Door middel van een lineaire functionaal, losgelaten op de oplossing

φ voor het homogenene abstracte Cauchy probleem, komen we tot een oplossing u voor de Volterra vergelijking.
In geval f niet identiek nul is, blijkt het homogene abstracte Cauchy probleem niet bruikbaar te zijn als halfgroep benadering. In plaats daarvan denken we aan een inhomogeneen abstract Cauchy probleem in een grotere ruimte dan H. Voor dit probleem bepalen we een oplossing in een geschikte interpolatieruimte $(H, \mathrm{D}(A))_{\eta,2}$ met $\eta \in (0,1)$, die we benutten om een oplossing voor de Volterra vergelijking te construeren.
Eén aspect in onze halfgroep benadering is niet onmiddelijk duidelijk. Een oplossing φ voor het homogene abstracte Cauchy probleem is een equivalentieklasse in de Hilbert ruimte, maar we zullen een versie van de equivalentieklasse nodig hebben om te bewijzen dat de geconstrueerde functie u de oplossing is voor de Volterra vergelijking. Dit probleem wordt behandeld in Hoofdstuk 2.
In Hoofdstuk 4 beschouwen we drie verschillende typen Volterra integrodifferentiaalvergelijkingen in een separabele Hilbert ruimte X. Paragraaf 4.1 beschrijft de X-waardige variant van de scalaire lineaire Volterra vergelijking. In Paragraaf 4.2 behandelen we de semilineaire X-waardige Volterra vergelijking

$$\begin{aligned} \frac{\mathrm{d}}{\mathrm{d}t} \int_{-\infty}^{t} a(t-s)u(s)\,\mathrm{d}s &= f(t, u(t)), \quad t > 0, \\ u(t) &= u_0(t), \qquad\quad t \leq 0. \end{aligned}$$

Paragraaf 4.3 gaat over de volgende lineaire X-waardige Volterra vergelijking van de eerste soort:

$$\int_0^t a(t-s)u(s)\,\mathrm{d}s = h(t), \quad t \geq 0.$$

Het idee om dit probleem te bestuderen komt voort uit de toepassing van een stochastische verstoring van de lineaire Volterra vergelijking in Hoofdstuk 5. In dit hoofdstuk beschouwen we de stochastische lineaire Volterra integrodifferentiaalvergelijking

$$\begin{aligned} \frac{\mathrm{d}}{\mathrm{d}t} \int_{-\infty}^{t} a(t-s)U(s)\,\mathrm{d}s &= f(t) + \sigma \dot{B}(t), \quad t > 0, \\ U(t) &= u_0(t), \qquad\qquad t \leq 0, \end{aligned}$$

waar $\{B(t)\}_{t\geq 0}$ standaard Brownse beweging is.

K.W. Homan,
Delft, februari 2003

Curriculum vitae

Krista Homan was born in Groningen on September 7, 1973. In 1991 she finished secondary school at the Praedinius Gymnasium Groningen and started her studies in Mathematics at the University of Groningen. The academic year 1994-1995 she spent at King's College London, being a keen member of its Mountaineering Club. Her graduation research was concerned with Dynamical Systems and was carried out at the Royal Dutch Meteorological Institute (KNMI). Under the supervision of prof.dr. H.W. Broer she studied behavioral changes in a climate model as a result of different values of parameters. With her thesis 'Routes to chaos in the Lorenz-84 atmospheric model' she obtained her M.Sc. degree in 1997.
In 1998 Krista was employed by Statistics Netherlands (CBS) as a project employee at the Population Department.
In September 1998 she started her Ph.D. research in Functional Analysis at the Delft University of Technology under the supervision of prof.dr. Ph. Clément. Part of the research was done in close cooperation with dr. G.W. Desch from the Karl-Franzens-Universität Graz in Austria. As a Ph.D. student Krista attended two summer schools at the Université de Besançon in France. For four years she has taught Analysis and Linear Algebra to freshers in Mechanical Engineering and Industrial Design. Also she was a member of the first-aid team of the Faculty of Electrical Engineering, Mathematics, and Computer Science in Delft. Finally, she is proud to have biked 20.000 kilometers between her home in The Hague and the Delft University of Technology.